Lecture Notes in Biomathematics

Managing Editor: S. Levin

82

Andrej Yu. Yakovlev
Nikolaj M. Yanev

Transient Processes in Cell Proliferation Kinetics

Springer-Verlag
Berlin Heidelberg New York London Paris Tokyo Hong Kong

Authors

Andrej Yu. Yakovlev
Leningrad Polytechnical Institute
Polytechnicheskaya ul., 29
Leningrad 195 251, USSR

Nikolaj M. Yanev
Institute of Mathematics
Bulgarian Academy of Sciences
8 Acad. G. Bonchev str.
1113 Sofia, Bulgaria

Translator

B.I. Grudinko
1st Leningrad
Medical Institute

Mathematics Subject Classification (1980): 60J85, 68J10, 62P10

ISBN 3-540-51831-2 Springer-Verlag Berlin Heidelberg New York Tokyo
ISBN 0-387-51831-2 Springer-Verlag New York Heidelberg Berlin Tokyo

Printing and binding: Beltz Offsetdruck, Hemsbach/Bergstr.
2146/3140-543210 – Printed on acid-free paper

In memory of our fathers

Yurij YAKOVLEV and Michail YANEV

TABLE OF CONTENTS

INTRODUCTION

A mathematician who has taken the romantic decision to devote himself to biology will doubtlessly look upon cell kinetics as the most simple and natural field of application for his knowledge and skills. Indeed, the thesaurus he is to master is not so complicated as, say, in molecular biology, the structural elements of the system, i.e. cells, have been segregated by Nature itself, simple considerations of balance may be used for deducing basic equations, and numerous analogies in other areas of science also add to one's confidence. Generally speaking, this superficial impression is correct, as evidenced by the very great number of theoretical studies on population kinetics, unmatched in other branches of mathematical biology. This, however, does not mean that mathematical theory of cell systems has traversed in its development a pathway free of difficulties or errors. The seeming ease of formalizing the phenomena of cell kinetics not infrequently led to the appearance of mathematical models lacking in adequacy or effectiveness from the viewpoint of applications. As in any other domain of science, mathematical theory of cell systems has its own intrinsic logic of development which, however, depends in large measure on the progress in experimental biology. Thus, during a fairly long period running into decades activities in that sphere were centered on devising its own specific approaches necessitated by new objectives in the experimental in vivo and in vitro investigation of cell population kinetics in different tissues.

There are at present quite a large variety of tools for experimental research in cell kinetics. The method that has received the widest acceptance is radioautography [2,6,9-13, 59] and its combinations with quantitative cytophotometry and time-lapse cinemicrography [11, 21 28, 50, 59], the latter being the only direct technique for measuring the duration of the mitotic cycle. By means of that technique information of primary importance for the theory of cell systems has been obtained on the specific distribution of generation time [32, 33, 36, 49, 50, 52] and duration of mitosis [53] for different types of cells in vitro. A semi-automatic system of processing films using a computer has been proposed [58] which extends considerably the potentialities of time-lapse cinemicrography as a method of studying individual behaviour of cells in a culture.

As a kind of alternative to radioautographic approach to cell kinetics studies a method was proposed named the BISACK system [56], based on introducing bromdeoxyuridine (BUdR) into DNA of actively proliferating cells. In the BISACK system doses of BUdR are used which induce no inhibition of DNA replication, and by means of differential fluorescent staining of chromosomes cells are revealed which appear in the metaphase for the first, second and third time during the period of observation. The method also enables determination of the total fraction of cells replicating DNA in the presence of BUdR . The BISACK system has been successfully employed in studying regularities in the kinetics of human skin fibroblasts in vitro [48], peripheral human blood lymphocytes stimulated with phytohemagglutinin (PHA) [54 - 56] and bone marrow cells of the rat [47,48]. It is worth while to compare results obtained by means of the BISACK system with the findings of radioautographic experiment. The principal difficulty involved in such a comparison lies in the fact that special methods must be used in analyzing radioautographic data which enable evaluation of cell kinetics not in the traditional terms of mitotic cycle phase durations but by determining fluxes of cells entering the cycle phases under study during a specified interval. In the present monograph we review such a method based on the introduction of the so-called q-index which characterizes the integral flux of cells into a given phase of the cell cycle. We have used this method, in combination with the algorithm of labelled mitoses curve analysis, also described in the monograph (see Chapter IV), in investigating the kinetics of PHA-induced proliferation of human blood lymphocytes, and our results are not at variance with those obtained by means of the BISACK system as reported in reference [64].

A particularly effective tool for studying cell kinetics is the method of flow cytofluorometry which enables a very high-speed (dozens of thousands of cells per minute) reproduction of histograms of DNA content of cells [27, 39, 46, 60 ,62]. Its limitation is the applicability only to cell suspensions, however, procedures for isolating cells from different tissues, solid tumors included, have attained a high degree of efficiency. No doubt, flow cytofluorometry holds much promise for the future, as regards not only scientific research but also practical medicine [23,42,57]. Gray [20] developed a method for analyzing distributions of DNA content of cells on the basis of the model of multiphase birth-death process. Using his method, Gray demonstrated a good agreement between the results obtained by the techniques of flow microfluorometry and radioautography. However, comparison of the techniques undertaken by other authors [37] has shown that radioautography may yield underestimated values for the fraction of cells in the S-phase of the mitotic cycle, presumably, due to the low radioactive label content in the slowly DNA-synthesizing cells of the population. Comparison of the two methods will also be found in reference [60]. The authors have created a computer model of rat spermatogenesis based on autoradiographic studies of cell cycle phase durations for each germ-cell type. The data calculated by means of the model and experimental flow cytometry findings have shown satisfactory agreement. Unfortunately, so far very few works have come out demonstrating combined application of all present-day methods for investigation of cell population kinetics.

With the accumulation of experimental material on cell kinetics in embryonal and definitive tissues, starting from the pioneer research by Howard and Pelc [25] and Quastler and Sherman [41] there has been a growing need for recruiting adequate mathematical apparatus for analysis of available results.

At present various approaches have satisfactorily been developed to the analysis of the following experimental data: (1) indices of labelled cells, using pulse [1,31] or continuous [1,26,43,44,64,66] labelling with ^{3}H-thymidine; (2) labelled mitoses curves with pulse [22, 24, 26,34,43,64] and continuous [44] labelling; (3) experimental evidence with double labelling [45,51]; (4) kinetic indices variation curves with the blocking of cell cycle processes [20, 26, 63]; (5) DNA synthesis and mitotic activity diurnal rhythm curves [22,30,35,65]; (6) distributions of grain counts in radioautographs and label dilution dynamics [14-16,51,69]; (7) flow microfluorometry data [3, 4, 7, 17-20,29,68,70]; (8) cell death characteristics [26].

Many of the methods of applied kinetic analysis are based upon probabilistic models of biologic population dynamics which, in turn, represent modifications of comprehensively studied models of stochastic processes of certain structures: various types of Markovian processes, age-dependent branching processes and renewal processes. Some of such models will be dealt with in this book. Attempts have been made to formalize cell kinetics on the basis of the theory of stochastic integral equations [40]. The fruitfulness of the stochastic approach has been vividly demonstrated in a recent work dealing with plant cell population growth [8]. There is, indeed, a vast literature on deterministic models of cell kinetics.

It is particularly important to single out intermediate type models which may be arbitrarily called semi-stochastic. Such models contain random variables side by side with deterministic parameters or functions. Besides, the following situation is of special interest: on the one hand, some parameters of a model (for instance, durations of cell cycle phases) are assumed to be random and, on the other, it is only the behaviour of the mathematical expectations of the principal variables (e.g., number of cells in a population or age distribution) that is investigated. We shall call such models semi-stochastic as well though a certain artificiality of the classification is evident. The notion of semi-stochastic model is similar to Nåsel's "hybrid model" introduced in his book [38].

Semi-stochastic models, given in this book quite an ample space, have a fairly wide range of applicability as regards description of cell kinetics phenomena and, at the same time, they are simple enough to serve as the basis for developing applied methods for analyzing experimental findings. In the past few years research associated with the development of stochastic simulation models of cell proliferation kinetics has been gaining in importance both in theoretical and applied respects [5,66].

The foregoing shows that, as regards present-day mathematical biology,there is no lack of models proposed for description and analysis of cell population kinetics.Much progress has also been made in developing experimental methods for exploring the structure of the cell cycle and their related software. However, in our opinion it is the abundance and diversity of the proposed mathematical methods that hinder their introduction into experimental practice. Some of the models are unjustifiably intricate for practical realization while others, on the contrary, contain biologically unacceptable assumptions. The user needs a single working apparatus embodying a reasonable compromise between achievements of the mathematical theory of cell systems and actual requirements of biologic experimental methodology. It is from this standpoint that problems involved in the description of transient processes in cell kinetics are considered in this book.

The investigator comes across the unsteady nature of cell kinetics in the majority of practically interesting situations. The latter, first and foremost , are associated with systems with induced or stimulated cell proliferation. In the meantime, many experts on simulation of population dynamics seek to obtain asymptotic results or concentrate on interrelations between different characteristics of an already steady state.

The purpose of the present book is to fill the gap caused in the present-day literature by the lack of the attention to transient processes in cell kinetics. We use the term "transient process" in the sense adopted in the theory of dynamic systems, i.e. applying it to time variations in the characteristics of a system due to initial conditions. It should be emphasized that the term, as used in the theory of branching stochastic processes, has an altogether different meaning.

Consideration for the effect of transient processes calls for modification of the existing methods for applied kinetic analysis. We have attempted to substantiate certain ways of such modification and to demonstrate their effectiveness on real biologic material, i.e. in studying regular features of the process of the rat liver regeneration following partial hepatectomy. Wherever possible, we endeavoured to provide a full probabilistic description of cell kinetics, giving up it only when the difficulties seemed insurmountable. Incidentally, in some cases the difficulties were of psychological rather than mathematical origin: we could not bring ourselves to define a priori the probabilistic structure of a process in the absence of corresponding experimental evidence or conceptual considerations. In handling all the problems, we gave priority to possible application of a mathematical result to analysis of experimental observations over the mathematical aspect of the problem. This could not but affect the manner of exposition which, we believe, will be acceptable to a fairly wide circle of readers. Chapters III and IV compared to the first two require less mathematical knowledge on the part of the readership, and they can be perused independently. Chapter V deals with biological applications. One interested mainly in biological aspects of cell kinetics may well confine himself to reading that particular chapter.

As most authors using mathematical symbols, we encountered considerable difficulties in trying to adhere to uniform literal notation. We were not always successful even within the framework of a given chapter. However, new designations were introduced in such a way as to prevent possible misunderstanding. The formulas are numbered consecutively within each chapter.

The volume structure.

Chapter I presents a brief introduction to the theory of stochastic branching processes. Purely auxiliary in concept, it aims at providing the reader with particulars on basic models (Markovian, age-dependent, multitype) of branching processes essential for perusal of subsequent chapters. In addition, it contains some references to attempts that have been made to resort to the theory of stochastic branching processes in resolving cell biology problems.

Chapter II is devoted to the application of the theory of branching age-dependent stochastic processes to research in systems with induced cell proliferation. Probabilistic description is given of cell generations in populations induced to proliferate and of certain processes relevant to analysis of data from radioautographic study of the kinetics of induced cell transition to DNA synthesis. The problems of theoretical description of grain count distribution within a proliferating cell population is considered as well as its applications to analysis of experimental radioautographic evidence. Certain possibilities are discussed of employing such a model in formalizing radiobiological effects.

Chapter III is concerned with development of basic equations that enable description of unsteady kinetics of transitive cell populations within semi-stochastic framework. Various ways of constructing transient processes for different states of cell proliferation kinetics are specifically dealt with. A mathematical model representing the blocking of cells in the mitotic cycle, including its application to estimating temporal cycle phase parameters, is examined. Consideration is given to a kinetic index (q - index) that characterizes the integral cell flow into a cycle phase, and a method is proposed for constructing it from experimental radioautographic data. Theoretical principles of using q-index in studying the kinetics of induced cell proliferation are discussed. The subject-matter of the chapter demonstrates how a combination of mathematical and experimental means can facilitate estimation of unobserved parameters of a model.

Chapter IV deals with mathematical modelling of labelled mitoses curve for the conditions of unsteady cell kinetics. The existing approaches to analyzing labelled mitoses curves are covered, as well as the limits of their applicability and the ways of generalizing them for the case of unsteady behaviour of cell populations. A procedure is proposed for constructing theoretical description of a fraction of labelled mitoses using q-index, and its specific applications under different states of cell proliferation are reviewed. The effect of transient processes on the form of labelled mitoses curve is investigated.

Chapter V illustrates potentialities of the applied methods developed in the book for exploring cell kinetics, using a specific example of the regenerating rat liver. Based on the literature, present-day concepts are defined of the temporal organization of hepatocyte transition to DNA synthesis and mitosis following partial hepatectomy. The phenomenon of "dynamic replacement of hepatocytes" in the regenerating liver revealed by the methods devised for studying induced cell proliferation kinetics is described, and its implication in maintenance of specialized functions of the liver during its regeneration after partial resection of hepatic parenchyma is specified. Results of interpreting data from cell culture studies are presented for comparison. A simple mathematical model is used for explaining regularities of proliferative response of hepatocytes to partial hepatectomy of dissimilar extent.

We wish to thank our numerous friends and colleagues who took part in the discussion of various problems touched upon in the book and, especially, Dr. B.I.Grudinko who undertook the arduous work of translating the manuscript from Russian into English. The technical assistance of G.M.Pukhova is greatly acknowledged.

REFERENCES

1. Appleton, D., Morley, A.R. and Wright, N.A. Cell proliferaition in the castrate mouse seminal vesicle in response to testosterone propionate.II.Theoretical considerations, Cell Tiss.Kinet., 6, 247-258, 1973.
2. Baserga, R. and Malamud, D. Autoradiography. Technique and applications, Harper and Row Publ., New York, 1969.
3. Bertuzzi, A., Gandolfi,A., Germani, A. and Viltelli, R. Estimation of cell DNA synthesis rate of cultured cells from flow cytometric data, Cytometry, 5, 619-628,1984.
4. Bertuzzi, A., Gandolfi, A. and Vitelli, R. A regularization procedure for estimating cell kinetic parameters from flow-cytometry data, Math. Biosci., 82, 63-85, 1986.
5. Biomathematics and Cell Kinetics, Rotenberg, M. ed., Elsevier/ North-Holland Biomedical Press, Amsterdam, New York, Oxford, 1981.
6. Bisconte, J.C. Kinetic analysis of cellular populations by means of the quantitative radioautography, Internat.Rev. Cytol., 57, 75-126, 1979.
7. Dean, P. and Jett, J. Mathematical analysis of DNA distributions derived from microfluorometry, J. Cell Biol., 40,523-527, 1974.
8. de Gunst, M. A random model for plant cell population growth, Doctors Thesis, University of Leiden, 1988.
9. Dörmer, P. Photometric methods in quantitative autoradiography, In.:Microautoradiography and electron probe analysis, Lüttge, U.(ed.), Springer, 7-48, 1972.
10. Dörmer, P. Quantitative autoradiography at the cellular level, Molec. Biol., Biochem., Biophys., 14, 347-393, 1973.
11. Dörmer, P., Brinkmann,W., Born, R. and Steel, G.G. Rate and time of DNA synthesis in individual Chinese hamster cells, Cell Tiss. Kinet., 8, 339-412, 1975.
12. Epifanova,O.I. and Terskikh, V.V. Radioautography in cell cycle research, Nauka, Moscow, 1969(In Russian).

13.Epifanova, O.I., Terskikh, V.V. and Zakharov, A.F. Radioautography, High School, Moscow, 1977(In Russian).
14.Fried, J. Correction for threshold error in the determination of generation times by the grain-count halving method, J.Theoret.Biol., 108-120, 1969.
15.Fried,J. A mathematical model to aid in the interpretation of radioactive tracer data from proliferating cell populations, Math. Biosci., 8, 379-396, 1970.
16.Fried,J. Proposal for the determination of generation time variability and dormancy of proliferating cell populations by a modification of the grain-count halving method, J.Theoret. Biol., 34, 533-555. 1972.
17.Fried,J. Method for the quantitative evaluation of data from flow microfluorometry, Comput. Biomed. Res.,9, 263-271, 1976.
18.Fried, J. and Mandel, M. Multi-user system for analysis of data from flow cytometry, Computer Programs in Biomedicine, 10, 218-230, 1979.
19.Fried, J., Perez, A. and Clarkson, B. Quantitative analysis of cell cycle progression of synchronous cells by flow cytometry Exper. Cell Res., 126, 63-74, 1980.
20.Gray, J.W. Cell-cycle analysis of perturbed cell populations: computer simulation of sequential DNA distributions, Cell Tiss. Kinet.,9, 499-516, 1976.
21.Gray, J.W., Carver, J.H., George, Y.S. and Mendelsohn, M.L. Rapid cell cycle analysis by measurement of the radioactivity per cell in a narrow window in S phase (RCS), Cell Tiss. Kinet., 10, 97-109, 1977.
22.Guiquet, M., Klein, B. and Valleron, A.J. Diurnal variation and the analysis of percent labelled mitoses curves, In: Biomathematics and Cell Kinetics, Elsevier /North-Holland Biomed. Press, Amsterdam, 191-198, 1978.
23.Harmon, J.M., Norman, M.R., Fowlkes, B.J., Thompson, E.B., Dexamethasone induces irreversible G_1/S arrest and death of a human lymphoid cell line, J.Cell. Physiol., 98, 267-278, 1979.
24.Hartmann, N.R., Gilbert, C.M., Jansson, B., Macdonald,P.D.M., Steel, G.G. and Valleron, A.J. A comparison of computer methods for the analysis of fraction labelled mitoses curves, Cell Tiss.Kinet., 8,119-124, 1975.
25.Howard, A. and Pelc,S.R. Synthesis of deoxyribonucleic acid in normal and irradiated cells and its relation to chromosome breakage, Heredity, Suppl., 6, 261-273, 1953.
26.Jagers, P. Branching processes with biological applications, Wiley, New York, 1975.
27.Kamentsky, L. Cytology automation, In: Advances in Biophys. and Med.Phys., Academic Press,New York, 83-142, 1973.
28.Killander, D. and Zetterberg, A. Quantitative cytochemical studies on interphase growth, Exper.Cell Res., 38, 272-284, 1965.
29.Kim, M. and Shin, K.G. Estimation of cell kinetic parameters from flow microfluorometry, Math. Biosci., 38, 77-89,1978.
30.Klein, B. and Valleron, A.J. Mathematical modelling of cell cycle and chronobiology: preliminary results, Biomedicine, 23,214-217, 1975.
31.Koschel, K.W., Hodgson, G.S. and Radley, J.M. Characteristics of the isoprenaline stimulated proliferative response of

rat submaxillary gland,Cell Tiss. Kinet., 9, 157-165, 1976.
32.Kubitschek,H.E. Normal distribution of cell generation rate, Exper. Cell Res., 26, 439-450, 1962.
33.Kubitschek, H.E. The distribution of cell generation times, Cell Tiss. Kinet., 4, 113-122, 1971.
34.Macdonald, P.D.M. Statistical inference from the fraction labelled mitoses curve, Biometrika, 57, 489-503, 1970.
35.Macdodnald, P.D.M. Measuring circadian rhythms in cell populations, In: The Mathematical Theory of the Dynamics of Biological Populations II, Academic Press, London, 1981.
36.Marshall, W.H., Valentine, F.T. an Lawrence, H.S. Cellular immunity in vitro. Clonal proliferation of antigen-stimulated lymphocytes, J.Exper.Med., 130, 327-342, 1969.
37.Møller, U. and Larsen. J.K. The circadian variations in the epithelial growth of the hamster cheek pouch: quantitative analysis of DNA distributions, Cell Tiss. Kinet., 11, 405-413, 1978.
38.Nåsell,I. Hybrid models of tropical infections, Lecture Notes in Biomathematics, Springer-Verlag, Berlin, Heidelberg, New York,Tokyo,1985.
39.Nicolini, C., Kendall, F., Baserga, R., Dessaive, C., Clarkson, B. and Fried, J. The G_0-G_1 transition of WI 38 cells. I. Laser flow microfluorimetric studies, Exper. Cell Res., 106, 111-118, 1977.
40.Padgett, W.J. and Tsokos, C.P. A new stochastic formulation of a population growth problem, Math.Bioscie., 17, 105-120, 1973.
41.Quastler, H. and Sherman, F.H. Cell population kinetics in the intestinal epithelium of the mouse, Exper. Cell Res., 17, 429-438, 1959.
42.Rutgers, D.H., Niessen. D.P.P. and Van der Linden, P.M. Cell kinetics of hypoxic cells in a murine tumour in vivo: flow cytometric determination of the radiation-induced blockage of cell cycle progression, Cell Tiss. Kinet., 20, 37-42, 1987.
43.Scheufens, E.E. and Hartmann, N.R. Use of gamma distributed transit times and the Laplace transform method in theoretical cell kinetics, J. Theor. Biol., 37,531-543, 1972.
44.Schotz, W.E. Continuous labelling indices: CLI(t) and CLM(t), J.Theoret. Biol.,34, 29-46, 1972.
45.Schotz, W.E. Double label estimation of the mean duration of the S-phase, J. Theor. Biol., 46, 353-368, 1974.
46.Shackney, S.E., Erickson, B.W. and Skramstad, K.S. The T-lymphocyte as a diploid reference standard for flow cytometry, Cancer Res., 39, 4418-4422, 1979.
47.Schneider, E.L., Sternberg, H. and Tice, R.R. In vivo analysis of cellular replication, Proc. Nat. Acad. Sci. USA, 74, 2041-2044, 1977.
48.Schneider, E.L., Sternberg, H., Tice, R.R. et al. Cellular replication and aging, Mechanisms of Ageing and Development, 9, 313-324, 1979.
49.Shields, R. and Smith, J.A. Cells regulate their proliferation through alterations in transition probability, J. Cell. Physiol., 91, 345-356, 1977.
50.Sisken,I.E. and Morasca, L. Intrapopulation kinetics of the mitotic cycle, J. Cell Biol., 25, 179-189, 1965.
51.Skagen, D.W. and Morkrid, L. An approach to the theory of quantitative and double label autoradiography, J.Theoret. Biol.,70, 185-197, 1978.
52.Smith, J.A. and Martin, L. Do cells cycle? Proc. Nat. Acad.

Sci. USA, 70, 1263-1267, 1973.

53. Rao, P.N. and Engelberg, J. Mitotic duration and its variability in relation to temperature in HeLa cells, Exper. Cell Res., 52, 198-208, 1968.

54. Tice, R., Schneider, E.L. and Rary, J.M. The utilization of bromodeoxyuridine incorporation into DNA for the analysis of cellular kinetics, Exper.Cell Res., 102, 232-236, 1976.

55. Tice, R., Schneider, E.L., Kram, D. and Thorne, P. Cytokinetic analysis of the impaired proliferative response of peripheral lymphocytes from aged humans to phytohemagglutinin, J. Exper. Med., 00, 1029-1041, 1979.

56. Tice, R., Thorne, P. and Schneider, E.L. BISACK analysis of the phytohemagglutinin-induced proliferation of human peripheral lymphocytes, Cell Tiss. Kinet., 12, 1-9, 1979.

57. Tobey, R.A. and Crissman, H.A. Use of flow microfluorometry in detailed analysis of effects of chemical agents on cell cycle progression, Cancer Res., 32, 2726, 1972.

58. Tolmach, A.P., Mitz, A.R., Rump, S.L., Pepper, M.L. and Tolmach, L.J. Computer-assisted analysis of time-lapse cinemicrographs of cultured cells, Computers and Biomed. Res., 11, 363-379, 1978.

59. Ucci, G., Riccardi, A., Dörmer, P. and Danova, M. Rate and time of DNA synthesis of human leukaemic blasts in bone marrow and peripheral blood, Cell Tiss. Kinet., 19, 429-436, 1986.

60. Van Dilla, M.A., Trujillo, T.T., Mullaney, P.F. et al. Cell microfluorometry: A method for rapid fluorescence measurement, Science, 163, 1213-1214, 1969.

61. Van Kroonenburgh, M.J., Van Gasteren, H.J., Beck, J.L. and Herman, C.J. A computer model of spermatogenesis in the rat; correlation with flow cytometric data based on autoradiographic cell-cycle properties, Cell Tiss. Kinet., 19, 171-177, 1986.

62. Wilson, G.D., Soranson, J.A. and Lewis, A.A. Cell kinetics of mouse kidney using bromodeoxyuridine incorporation and flow cytometry: preparation and staining, Cell Tiss. Kinet., 20, 125-133, 1987.

63. Yakovlev, A.Yu. On the simulation of mitotic block induced by irradiation, Cytology, 15, 616-619, 1973 (In Russian).

64. Yakovlev, A.Yu. Kinetics of proliferative processes induced by phytohemagglutinin in irradiated lymphocytes, Radiobiology, 23, 449-453, 1983 (In Russian).

65. Yakovlev, A.Yu., Lepekhin, A.F. and Malinin, A.M. The labeled mitoses curve in different states of cell proliferation kinetics . V. The influence of diurnal rhythm of cell proliferation on the shape of the labeled mitoses curve, Cytology, 20, 630-635, 1978 (In Russian).

66. Yakovlev, A.Yu. and Zorin, A.V. Computer simulation in cell radiobiology, Springer-Verlag, Berlin, Heidelberg, New York, 1988.

67. Yakovlev, A.Yu., Zorin, A.V., and Isanin, N.A. The kinetic analysis of induced cell proliferation, J. Theoret.Biol.,64, 1-25, 1977.

68. Yanagisawa, M., Dolbeare, F., Todoroki, T. and Gray, J.W. Cell cycle analysis using numerical simulation of bivariate DNA/bromodeoxyuridine distributions, Cytometry,6,550-562, 1985.

69. Yanev, N.M. and Yakovlev, A.Yu. On the distribution of marks over a proliferating cell population obeying the Bellman-

Harris branching process, Math. Biosci., 75, 159-173,1985.
70.Zietz, S. FP_1 Analysis.I. Theoretical outline of a new method to analyze time sequences of DNA histograms, Cell Tiss.Kinet.,13, 461-471, 1980.

I. SOME POINTS OF THE THEORY OF BRANCHING STOCHASTIC PROCESSES

1.1. Introduction

This chapter outlines (with no proof presented) certain points of the theory of branching stochastic processes which will be of use in reading Chapter II. In addition, we have included here some theorems on the asymptotic behaviour of the Bellman-Harris process as well as some other results most frequently utilized in applications to cell population kinetics. It is presumed that the reader is familiar with the fundamentals of the probability theory in its present-day form. For this reason no further explanations will be given in connection with such terms as probability space, random variable (process,field), distribution function or generating function. However, all concepts and definitions peculiar to the theory of branching processes are given in a form that makes it unnecessary for the reader to refer to corresponding monographs or manuals. A number of excellent books are available on the theory of branching processes. Monographs by Athreya and Ney [2],Harris [4], Mode [9], Sevastyanov [12], Jagers [6] and Assmusen and Hering [1] deserve a special mention. Those were the sources used in the brief review that follows.

1.2. The Galton-Watson Process

A process $\{\mu_t\}$ is called a Markov process if its "future" depends on the "past" only via the "present". In terms of transition probabilities this property is expressed as

$$\mathbb{P}\{\mu_{n+t}=j\,|\,\mu_0=i_0,\mu_1=i_1,\dots,\mu_{n-1}=i_{n-1},\mu_n=i\}=$$
$$\mathbb{P}\{\mu_{n+t}=j\,|\,\mu_n=i\} = P\{\mu_t=j\ |\mu_0=i\} = p_{ij}(t). \tag{1}$$

The last two equalities in (1) indicate that only homogeneous (in time) Markov processes are implied.

The Galton-Watson branching process μ_t may be introduced in the following constructive way. Let integer-valued non-negative random variables $\xi_i(t)$, $i=1,2,\ldots$, $t=0,1,2,\ldots$, assumed to be independent and identically distributed, be defined on a probability space $(\Omega,\mathcal{F},\mathbb{P})$ and

$$\mathbb{P}\{\xi_i(t)=k\} = P_k, \quad \sum_{k=0}^{\infty} P_k = 1.$$

Then for the Galton-Watson process we assume

$$\mu_{t+1} = \begin{cases} \sum_{i=1}^{\mu_t} \xi_i(t), & \text{if } \mu_t>0, \\ 0, & \text{if } \mu_t=0. \end{cases} \tag{2}$$

It is generally assumed that $\mu_0=1$ since the results obtained under that condition can readily be extended (as we shall see later) to a more general case when μ_0 is an integer-valued non-negative random variable independent of $\{\xi_i(t)\}$. Therefore, hereinafter we shall in all cases consider that $\mu_0=1$.
If the parameter t is interpreted as the number of particle generation and the process states as the number of particles, then $\xi_i(t)$ will denote the size of the i-th particle's progeny existing in the t-generation. In that case μ_t is the total number of particles in the t-generation. The principal property of the Galton-Watson process expressed in (2) is the independence of each particle's evolution from the total number of particles existing in a given generation. It is precisely that property that is responsible for the extensive use of the powerful apparatus of generating functions in the theory of branching processes.

Let us call the numbers $P_k=\mathbb{P}\{\mu_1=k\}$ as individual probabilities. It will then be natural to call the function $\Phi(s)=\mathbb{E}s^{\mu_1}$ as an individual generating function. Individual characteristics completely determine all other characteristics of the μ_t process, including transition probabilities

$$p_n(t) = \mathbb{P}\{\mu_{t+\tau}=n \,|\, \mu_\tau=1\} = \mathbb{P}\{\mu_t=n\}$$

and generating functions corresponding to them

$$\Psi(t;s) = \sum_{n=0}^{\infty} p_n(t)s^n = \mathbb{E}s^{\mu_t} .$$

For the Galton-Watson processes property (1) follows immediately from the relation

$$\mu_{t+n} = \sum_{i=1}^{\mu_n} \mu_t^{(i)}$$

where $\{\mu_t^{(i)}\}$ are independent and identically distributed (with the same distribution as μ_t) random variables independent of μ_n , and t and n are any integer non-negative numbers. Thus, the Galton-Watson process is a particular case of the Markov process taking on values from the phase space $N = \{0,1,2,...\}$. Such a process is usually called a Markov chain, with discrete time if the parameter t takes on values from the set N, or continuous time if $t \in R_+^1$.

Hereinafter the set T will be taken to mean either the set N or the interval $[0,\infty)$. Then the transition probabilities $p_{ij}(t)$ of the Markov chain will meet the following conditions:

(a) $p_{ij}(t) \geq 0$ at all $i,j \in N$, $t \in T$
(non-negativity condition);

(b) $\sum_{j=0}^{\infty} p_{ij}(t)=1$ at any $i \in N$, $t \in T$
(normalizing condition);

(c) $p_{ij}(t+u)=\sum_{k=0}^{\infty} p_{ik}(t)p_{kj}(u)$ for any $i,j \in N$; $u,t \in T$
(Markovian condition);

(d) $p_{ij}(0)=\delta_{ij}=\begin{cases} 1, & i=j \\ 0, & i\neq j \end{cases}$, (initial condition).

In case of continuous time the following continuity condition is usually imposed too:

(e) $\lim_{t\to 0+} p_{ii}(t)=1$

Then from the conditions (a),(b) and (e) it follows at once that for any i uniformly for all $j\neq i$ $\lim_{t\to 0+} p_{ij}(t)=0$,while from the condition (c) stems the continuity of all transition probabilities $p_{ij}(t)$ at any $t\geq 0$.

Note that those conditions determine completely a Markov chain. For instance, in case of discrete time $t=0,1,2,...$ it is sufficient to define transition probabilities for one step: $p_{ij}=p_{ij}(1)$.

The Markov chain on N will be a branching process if the transition probabilities $p_{ij}(t)$ also satisfy, besides the conditions (a)- (e), the branching condition:

$$(f)\quad p_{ij}(t) = p_{1j}^{*i}(t) = \sum_{j_1+j_2+\ldots+j_i=j} p_{1j_1}(t)p_{1j_2}(t)\ldots p_{1j_i}(t)$$

i.e. $p_{ij}(t)$ is the i-fold convolution of the distribution $p_{1j}(t)$. With $i=0$ the condition (f) takes the form $p_{0j}(t)=p_{1j}^{*0}(t)=\delta_{0j}$.

In terms of generating functions the condition (f) implies that

$$\Psi_i(t;s) = \sum_{j=0}^{\infty} p_{ij}(t)s^j = \mathbb{E}\{s^{\mu_t}|\mu_0=i\} = \sum_{j=0}^{\infty} p_{1j}^{*i}(t)s^j =$$

$$(\sum_{j=0}^{\infty} p_{1j}(t)s^j)^i = (\mathbb{E}\{s^{\mu_t}|\mu_0=1\})^i = \Psi^i(t;s) . \tag{3}$$

Similarly, by multiplying both sides in (c) by s^j and summing over j, we have

$$\Psi_i(t+u;s) = \sum_{j=0}^{\infty} p_{ij}(t+u)s^j = \sum_{j=0}^{\infty}\sum_{k=0}^{\infty} p_{ik}(t)p_{kj}(u)s^j =$$

$$\sum_{k=0}^{\infty} p_{ik}(t)\sum_{j=0}^{\infty} p_{kj}(u)s^j = \sum_{k=0}^{\infty} p_{ik}(t)\Psi^k(u;s) = \Psi_i(t;\Psi(u;s)),$$

whence for the generating function of the Galton-Watson process come the basic functional equation

$$\Psi(t+u;s) = \Psi(t;\Psi(u;s)) \tag{4}$$

for any $t,\ u \geq 0$, $|s| \leq 1$. The initial condition for (4) has the form: $\Psi(0;s)=s$.

Thus, the branching and Markovian conditions are equivalent to equations (3) and (4).

1.3. The Bellman-Harris Process

Present-day mathematical models of temporal organization of the cell cycle are usually based on considering the random time a cell spends in a given population (cycle phase) T which is characterized by a certain probability distribution $G(x)=\mathbb{P}\{T \leq x\}$ defined on R^1_+ . As any non-decreasing (continuous on the right) function $G(x)$ may be represented by the sum of the three components: $G(x)=a^2A(x)+b^2B(x)+c^2Z(x)$, where $A(x)$ is the absolutely continuous function, $B(x)$ is the step-function and $Z(x)$ is the singular component, i.e. the continuous function with a bounded total variation possessing a derivative which is zero almost everywhere. The coefficients a,b,c satisfy the condition: $a^2+b^2+ c^2= 1$. In most cases the $G(x)$ distribution is classified under type $A(x)$ or type $B(x)$. In so doing the notion of lattice distribution is distinguished, i.e. of a step function whose jumps are located at the points $k\Delta$ where k is the positive integer and Δ is the largest of the number for which

$$\sum_{k=0}^{\infty} [B(k\Delta+0) - B(k\Delta-0)] = 1 .$$

In application use is generally confined to the absolutely continuous component, representing $G(x)=A(x)$ as

$$G(x) = \int_0^x g(u)du ,$$

where $g(x)$ is the distribution density for T .In solving concrete problems, it is often assumed that $g(x)$ is not only integrable but continuous as well.

As regards the numerical parameters, of major interest are the first two moments of $G(x)$ distribution, i.e. the mean value

$$\bar{\tau} = \int_0^{\infty} xdG(x) = \int_0^{\infty} [1-G(x)]dx ,$$

and variance

$$\sigma^2 = \int_0^{\infty} [x - \bar{\tau}]^2 dG(x) .$$

One of the common problems of cell kinetics consists in obtaining estimates of these very parameters for different cell cycle phases.

Let us now consider a model of a branching process with continuous time and age-dependent transformations of particles (an age-dependent branching process) which is more complex than the Galton-Watson process and is not Markovian in the general case. It is the Bellman-Harris branching process model named so after its first investigators. In constructing the model, it is also assumed that particles (cells) undergo evolution independently of one another. As compared to the Galton-Watson process the new element consists in regarding every particle (cell) as having a random life-span T with the distribution function $G(x)$ and producing at the end of its life ν zero-aged particles. The random variable ν is characterized by the generating function

$$h(s) = \mathbb{E}s^{\nu} = \sum_{k=0}^{\infty} \mathbb{P}(\nu=k)s^k = \sum_{k=0}^{\infty} P_k s^k .$$

Realization of a particle's life time and the number of its direct progeny are referred to as the evolution of the particle.

Let $\mu(t)$ denote the number of particles existing at the instant $t \geq 0$. Introducing the familiar designations we can write

$$p_n(t) = \mathbb{P}\{\mu(t)=n|\mu(0)=1\} ,$$

$$\Psi(t;s) = \sum_{n=0}^{\infty} p_n(t)s^n = \mathbb{E}\{s^{\mu(t)}|\mu(0)=1\} .$$

Then the basic assumption of the independent evolutions of particles would imply that

$$\mathbb{E}\{s^{\mu(t)}|\mu(0)=m\} = \Psi^m(t;s) , \qquad (5)$$

if at the initial instant $t=0$ the particles are of zero age, which assumption will hold in the sequel unless otherwise is specified.

It is easy to see that in the general case the Bellman-Harris process is not Markovian since, knowing only the number of particles at the instant t (i.e. the states of the process $\mu(t)$), we cannot determine probabilities of the states $\mu(t+u)$, $u > 0$ because the particles existing at the instant t are of different age, i.e. the "future" of the process depends not only on the "present" but also on its "past". Therefore the basic functional equation (4) can no longer be used here. The generating function $\Psi(t;s)$ of the Bellman-Harris process satisfies the following non-linear integral equation

$$\Psi(t;s) = \int_0^t h(\Psi(t-u;s))dG(u)+s(1-G(t)) \tag{6}$$

with the initial condition $\Psi(0;s)=0$.

For the sake of the sequel it is expedient to give here the derivation of the equation.

First let us take the conditional expectation in the expression $\Psi(t;s)=\mathbb{E}(\mathbb{E}\{s^{\mu(t)}|(T,\nu)\})$ where (T,ν) is the initial particle's evolution.

If $T>t$, then $\mu(t)=1$, hence,

$$\mathbb{E}\{s^{\mu(t)}|T > t\} = s$$

If $T\le t$, then

$$\mu(t) = \sum_{i=1}^{\nu} \mu^{(i)}(t-T) ,$$

where $\{\mu^{(i)}(t-T)\}$ are independent (moreover, not depending on ν), identically distributed random variables.

Hence,

$$\mathbb{E}\{s^{\mu(t)}|(T \le t,\nu)\} = \Psi^{\nu}(t-T;s) .$$

Now, averaging over the distribution of (T,ν), we obtain

$$\Psi(t;s) = s(1-G(t)) + \int_0^t \sum_{k=0}^{\infty} \mathbb{P}\{\nu=k\}\Psi^k(t-u;s)dG(u) =$$

$$s(1-G(t)) + \int_0^t h(\Psi(t-u;s))dG(u) .$$

It may be proved that with $G(0)=0$ and $h'(1) < \infty$ the unique solution of equation (4) exists in the class of probabilistic generating functions. Harris [4] has shown that in this case the function $\mu(t)$ is bounded with probability one and equation (6) may be used for calculating the moments of $\mu(t)$.

Let us introduce designation $M_1(t) = \mathbb{E}\mu(t)$ and assume that $\eta = h'(1) < \infty$. Differentiating (6) with respect to s and assuming $s=1$ for the function $M_1(t)$ we obtain the integral equation

$$M_1(t)=1-G(t)+\eta\int_0^t M_1(t-u)dG(u), \tag{7}$$

which is sometimes referred to as the Harris-Bellman equation. At $\eta > 1$ the function $M_1(t)$ does not decrease, and at $\eta < 1$ does not increase with respect to t . If, in addition, $h''(1) < \infty$, then, differentiating (6) twice with respect to s and assuming s=1, it is also possible to obtain a corresponding equation for the second initial moment

$$M_2(t) = h''(1) \int_0^t (M_1(t-u))^2 dG(u) + \eta \int_0^t M_2(t-u)dG(u)+1-G(t), \tag{8}$$

where the function $M_1(t)$ is assumed to be known from (7). The variance $\mathbb{D}\mu(t)$ is determined by the formula $\mathbb{D}\mu(t)=M_2(t)-M_1^2(t)$.

For the product moment of the second order, i.e. $M_2(t,\tau)= = \mathbb{E}\{\mu(t)\mu(t+\tau)\}$, the following functional equation holds

$$M_2(t,\tau)=\eta \int_0^t M_2(t-u,\tau)dG(u)+h''(1) \int_0^t M_1(t-u)M_1(t+\tau-u)dG(u) +$$

$$\int_t^{t+\tau} M_1(t+\tau -u)dG(u)+1-G(t+\tau).$$

Equation (7) is a particular case of the so-called renewal

equation, i.e. a linear integral equation of the type

$$Y(t) = K(t) + \eta \int_0^t Y(t-u)dG(u) . \qquad (9)$$

Let us now introduce a designation for the one-sided convolution of the functions f and h

$$[f * h](t) = \int_0^t f(t-u)dh(u) .$$

Equation (9) has a unique solution $Y(t)$ which is non-negative and is representable in the form

$$Y(t)=K(t) + \sum_{n=1}^{\infty} \eta^n [K * G^{*n}](t) , \qquad (10)$$

where

$$G^{*1}(t)=G(t), \; G^{*(n+1)}=[G^{*n} * G](t) .$$

From formula (IO) we at once obtain solution of equation (7)

$$M_1(t)=\bar{G}(t) + \sum_{n=1}^{\infty} \eta^n [\bar{G} * G^{*n}](t) ,$$

where $\bar{G}(t) = 1-G(t)$. Similarly a solution may be obtained for equation (8) which is also an equation of the renewal type. Some properties of the renewal equation will be considered in Chapter III.

There are two important Markovian exceptions to the general non-Markovian Bellman-Harris model.

One is defined by the condition

$$G(x) = \begin{cases} 0, \text{ if } x \le 1 , \\ 1, \text{ if } x>1 . \end{cases}$$

In that case it follows from (6) that $\Psi(t;s)=h(\Psi(t-1;s))= =h_t(s)$, i.e. $\mu(t)$ is a Galton-Watson process with the individual generating function $h(s)$.

The other exception holds the case of

$$G(x) = \begin{cases} 0 , \text{ if } x \le 0 , \\ 1-e^{-\lambda x} , \text{ if } x>0, \end{cases}$$

where $\lambda>0$.

Here the Markovian character of the branching process $\mu(t)$

is due to the following exclusive property ("lack of memory") of the exponential distribution

$$G(x,y) = \mathbb{P}\{T-y \le x \,|\, T > y\} = G(x) . \tag{11}$$

Actually

$$G(x,y) = \frac{\mathbb{P}\{y<T \le x+y\}}{\mathbb{P}\{T>y\}} = \frac{G(x+y) - G(y)}{1 - G(y)} =$$

$$\frac{e^{-\lambda y}-e^{-\lambda(x+y)}}{e^{-\lambda y}} = 1 - e^{-\lambda x} = G(x) .$$

It follows from (11) that for a particle existing at the moment t the probability to survive for the time x does not depend on how long it has lived until the moment t, i.e. the "residual" life-time again has an exponential distribution with the same parameter λ. Then, it implies that for any $t,u \ge 0$

$$\mu(t+u) = \sum_{i=1}^{\mu(t)} \mu^{(i)}(u) , \tag{12}$$

where $\{\mu^{(i)}(u)\}$ are mutually independent, distributed identically with $\mu(u)$, random variables independent of $\mu(t)$.

As we know, from (12) follow the Markovian property and the basic functional equation

$$\Psi(t+u;s)=\Psi(t;\Psi(u;s)) . \tag{13}$$

On the other hand, from (6) we have

$$\Psi(t;s)= \int_0^t h(\Psi(t-u;s))\lambda e^{-\lambda u}du + se^{-\lambda t},$$

whence, differentiating with respect to t, we obtain

$$\frac{\partial\Psi(t;s)}{\partial t} = \lambda\, [h(\Psi(t;s))-\Psi(t;s)] .$$

Let us denote

$$f(s) = \lambda\, [h(s)-s] = \sum_{k=0}^{\infty} r_k s^k , \tag{14}$$

where

$$r_k=\lambda P_k \geq 0, k\neq 1; \; r_1=\lambda(P_1-1)<0; \; f(1)=\sum_{k=0}^{\infty} r_k=0 \; .$$

Thus, in the Markovian case the generating functions of the process $\Psi(t;s)$ satisfy the ordinary differential equation

$$\frac{\partial\Psi(t;s)}{\partial t} = f(\Psi(t;s))$$

with the initial condition $\Psi(0;s)=s$.

Let us try to ascertain the meaning of the infinitesimal characteristics $\{r_n\}$ which may be called transition probability densities.

Let us first calculate the probability of one particle transforming into n particles within the interval Δt in which no other transformation has taken place. It can be easily calculated that with $n\neq 1$ this probability is

$$(1-e^{-\lambda\Delta t})P_n=\lambda P_n\Delta t+o(\Delta t)=r_n\Delta t+o(\Delta t).$$

Similarly, the probability of one particle yielding within the time Δt exactly one particle, provided not more than one transformation takes place within the interval, is

$$e^{-\lambda\Delta t}+(1-e^{-\lambda\Delta t})P_1 = 1+\lambda(P_1-1)\Delta t+o(\Delta t)=1+r_1\Delta t+o(\Delta t)$$

On the other hand, the probability of more than one transformation occurring within the time Δt is $o(\Delta t)$. Indeed,

$$\begin{aligned}
&\mathbb{P}\{\text{more then one transformation in } \Delta t\} = \\
&1-\mathbb{P}\{0 \text{ transformations in } \Delta t\} - \\
&\mathbb{P}\{\text{exactly 1 transformation in } \Delta t\} = \\
&1-e^{-\lambda\Delta t}-(1-e^{-\lambda\Delta t})e^{-\lambda\Delta t}=o(\Delta t).
\end{aligned}$$

Thus, we have found that at $\Delta t \to 0$ the probabilities of the transition $p_n(\Delta t) = \mathbb{P}\{\mu(\Delta t)=n \mid \mu(0)=1\}$ may be presented in the following form

$$p_1(\Delta t)=1+r_1\Delta t+o(\Delta t) ,$$
$$p_n(\Delta t)=r_n\Delta t+o(\Delta t) , n\neq 1 . \qquad (15)$$

Bearing in mind that $p_1(0)=1$ and $p_n(0)=0$ at $n\neq 1$, we obtain from (15) that transition probabilities $p_n(t)$ are differentiable at zero point, their derivatives at zero are equal to transition probability densities.

1.4. Asymptotic Behaviour of the Bellman-Harris Process Characteristics

Asymptotic results reveal important properties of a branching process.

Specifically, the following result demonstrates asymptotically exponential growth of the expected size of a population when the generation coefficient (offspring mean) $\eta>1$. If the constant α , called the Malthusian population parameter, is defined as the positive root of the characteristic equation

$$\eta\int_0^\infty e^{-\alpha t}dG(t)=1$$

and if $G(x)$ is a non-lattice distribution, then holds the following asymptotic equality

$$M_1(t) \sim ce^{\alpha t}, t\to\infty , \qquad (16)$$

where

$$c = \frac{\eta-1}{\eta^2\alpha \int_0^\infty xe^{-\alpha x}dx} .$$

Result (16) may be strengthened if $G(x)$ possesses the density $g(x) \in L_\gamma(0,\infty)$ at $\gamma>1$. In that case

$$M_1(t)=ce^{\alpha t}[1+O(e^{-\varepsilon t})], t\to\infty , \varepsilon>0.$$

With $\eta=1$ the mean size of a population is apparently constant, whereas with $\eta < 1$ it may be shown that $\lim_{t\to\infty} M_1(t)=0$.

If $\eta>1$ and $h''(1)<\infty$ and the distribution $G(x)$ is non-lattice, then

$$M_2(t,\tau) = \frac{h''(1)c^2\int_0^\infty e^{-2\alpha u}dG(u)}{1-\eta\int_0^\infty e^{-2\alpha u}dG(u)}\ e^{\alpha\tau+2\alpha t}[1+o(1)]\ ,\ t\to\infty\ ,$$

where $\lim_{t\to\infty} o(1)=0$ uniformly with respect to τ, the values c and α being defined in the foregoing. The asymptotic results given above also have corresponding analogues for the case of a lattice distribution. More refined results concerning the asymptotic behaviour not only of moments but also of the random function $\mu(t)$ proper, are given in Harris' monograph [4].

A branching process is termed subcritical if $\eta<1$, supercritical if $\eta>1$, and critical if $\eta=1$, $h''(1)>0$. In the case of $\eta=1$, $h''(1)=0$ a branching process degenerates into a simple renewal process.

Let us introduce a point random process $N(a,t)$ determining the number of cells of an age not exceeding a at the instant t. The generating function $R(a,s,t)$ of the random process $N(a,t)$ satisfies the functional equation

$$R(a,s,t) = [1-G(t)][sJ(a-t)+1-J(a-t)]+\int_0^t h(R(a,s,t-u))dG(u),$$

where $J(t)$ is defined by the conditions: $J(t)=1$ at $t\geq 0$ and $J(t)=0$ at $t<0$. Hence, for the distribution of "expectations among ages" $M_1(a,t)=\mathbb{E}\{N(a,t)\}$ we have

$$M_1(a,t)=[1-G(t)]J(a-t)+\eta\int_0^t M_1(a,t-u)dG(u).$$

In the case of non-lattice $G(x)$ and $\eta>1$, with each a from any finite interval prevails the asymptotic equality

$$M_1(a,t)\sim ce^{\alpha t}B(a),\ t\to\infty\ .$$

The function $B(a)$, called the limiting age distribution, with any real η has the form

$$B(a) = \frac{\int_0^a e^{-\alpha t}[1-G(t)]dt}{\int_0^\infty e^{-\alpha t}[1-G(t)]dt} .$$

If a population does not become extinct, the limiting distribution $B(a)$ also mirrors the actual age distribution with $t \to \infty$. On this point Harris has proved:

THEOREM. Let $\eta>1$ and $h''(1)<\infty$ and the density $g(x) \in L_\gamma(0,\infty)$ of the distribution $G(x)$ existing at $\gamma>1$. Then

$$\mathbb{P}\left\{\lim_{t\to\infty} \frac{N(a,t)}{ce^{\alpha t}} = B(a)W;\ a \ge 0\right\} = 1,$$

where the random variable W which has an expected value equal to 1 and a positive variance may be characterized by the moments generating function $\Pi(s)$ which satisfies the functional equation

$$\Pi(s) = \int_0^\infty h\,[\Pi(se^{-\alpha x})]dG(x) , \quad \mathrm{Re}\ s \ge 0.$$

Studying a cell population stemming from some ancestor cell of y-age and proliferating by binary splitting, Nooney [10] thoroughly investigated the asymptotic behaviour of the first two moments of age distribution with due regard for the possible death of cells during the cycle. In constructing the functional equation for the generating function

$$R(s,a,y,t) = \sum_{k=0}^{\infty} \mathbb{P}\{N(a,y,t)=k\}s^k;\ t\ge 0,\ |s|\le 1, \tag{17}$$

the author proceeded from the following auxiliary probability characteristics of the process:

$p(y,t)=\frac{[G(y+t)-G(y)]}{[1-G(y)]}$ – probability that a cell of age y at the moment $t=0$ would divide producing two descendants not later than the moment t given that cell death in the meantime is excluded,

$q(y,t)=\frac{[Q(y+t)-Q(y)]}{[1-Q(y)]}$ – probability that a cell of age y at the moment $t=0$ would die not later than the moment t without undergoing division.

In the second expression $Q(a)$ is the conditional function of cell death time distribution, i.e. the probability of death at an age below or equal to a if the cell undergoes no division. Relying on the definitions of the functions $p(y,t)$ and $q(y,t)$ and confining himself to the case of $h(s)=s^2$, Nooney derived for generating function (17) of the process $N(a,y,t)$ the following functional equation

$$R(s,a,y,t)=q(y,t)[1-p(y,t)]+\int_0^t q(y,u)d_u p(y,u) +$$

$$[1-q(y,t)][1-p(y,t)]J(y+t-a)+s[1-q(y,t)][1-p(y,t)][1-J(y+t-a)] +$$

$$\int_0^t (R(s,a,0,t-u))^2[1-q(y,u)]d_u p(y,u) \quad . \qquad (18)$$

The expectation and variance are to be found by the customary formulas

$$M_1(a,y,t)=R'_s(1,a,y,t);$$

$$D(a,y,t)=R''_s(1,a,y,t)+M_1(a,y,t)-(M_1(a,y,t))^2.$$

Differentiating (18) with respect to s once and twice,

respectively, and assuming $s=1$, we formally obtain

$$M_1(a,y,t)=[1-q(y,t)][1-p(y,t)][1-J(y+t-a)] + 2\int_0^t M_1(a,0,t-u)[1-q(y,u)]d_u p(y,u) , \tag{19}$$

$$R''(1,a,y,t)=2\int_0^t \{(M_1(a,0,t-u))^2 + R''(1,a,0,t-u)\}[1-q(y,u)]d_u p(y,u) \tag{20}$$

Letting $V(a,y,t)$ represent the expression in the brace, we have instead of (20)

$$V(a,y,t)=(M_1(a,y,t))^2+2\int_0^t V(a,0,t-u)[1-q(y,u)]d_u p(y,u) . \tag{21}$$

To study the asymptotic properties of the first two moments of age distribution it is convenient to let $y=0$ and turn to auxiliary renewal equations of the type

$$M_1(a,0,t)=[1-Q(t)][1-G(t)][1-J(t-a)] + 2\int_0^t M_1(a,0,t-u)[1-Q(u)]dG(u) , \tag{22}$$

$$V(a,0,t)=(M_1(a,0,t))^2 + 2\int_0^t V(a,0,t-u)[1-Q(u)]dG(u) . \tag{23}$$

Applying different Tauberian type theorems to equations (22) and (23) Nooney then investigated the asymptotic behaviour of the moments using equations (19) and (21) supplemented with the formula for variance

$$D(a,y,t)=V(a,y,t)-2[M_1(a,y,t)]^2+M_1(a,y,t).$$

His final conclusions were formulated for a population with an arbitrary initial distribution $n(y,0)$ and in that form will be given here. Taking into account the independent evolution of individual cells, the author redetermined the mean and variance by means of expressions

$$\tilde{M}_1(a,t)=\int_0^\infty M_1(a,y,t)d_y n(y,0); \quad \tilde{D}(a,t)=\int_0^\infty D(a,y,t)d_y n(y,0);$$

and introduced the symbol

$$\beta(t) = 2\int_0^t [1-Q(u)]dG(u) .$$

Naturally, three cases: (a) $\beta(\infty)<1$, (b) $\beta(\infty)=1$ and (c) $\beta(\infty)>1$ can be distinguished. Nooney has demonstrated that under certain extra conditions contained in the formulations of Tauberian theorems in the case of (a) the functions $\tilde{M}_1(a,t)$ and $\tilde{D}(a,t)$ tend to zero as $t\to\infty$, in the case of (b) there are the finite limits $\lim_{t\to\infty} \tilde{M}_1(a,t)$ and $\lim_{t\to\infty} \tilde{D}(a,t)\ t^{-1}$ different from zero, and in the case of (c) there exist the limits: $\lim_{t\to\infty} \tilde{M}_1(a,t)\exp(\alpha' t)$ and $\lim_{t\to\infty} \tilde{D}(a,t)\exp(-2\alpha' t)$ (where α' is defined as the root of the characteristic equation $2\int_0^\infty e^{-\alpha' u}[1-Q(u)]dG(u)=1$) which are different from zero.

The above short summary of results from the theory of branching processes shows that a thoroughly studied mathematical model may be used for describing cell populations and that the difficulties that may arise will mainly center on whether such a model adequately fits experimental material. The work of Nedelman et al. [11] contains a long list of references to papers outlining results of using the mathematical apparatus of branching processes theory to elucidate the dynamic behaviour of many different kinds of cells: colony-forming blood cells growing in the spleens of irradiated mice, stem cells of hydra growing in tissue culture, fibroblasts, precursors of fat cells, tumor cell lines in culture and others. In his studies (see[8]) Kimmel developed a model of cell cycle based on multitype branching process in varying environment. The main purpose of that model was to describe the perturbations of leukemic cell population in the course of chemotherapy. Branching processes have proved a convenient tool in studying cell surface aggregation phenomena [8]. The work by Nedelman et al. [11] mentioned above, is devoted to employing an age-dependent, multitype branching process model in investigating the growth dynamics of mast cell colonies. We shall revert to that work in the next chapter. There are many generalizations of the type of model discussed here (specifically, the results of Crump and Mode [3] take into account correlation of cycle durations of sister cells). Generalizations of the initial model, however, often involve a considerably more elaborate mathematical apparatus, sometimes making analytical investigation impracticable.

In conclusion let us dwell on a very important generalization of the Bellman-Harris process proposed by Jagers [6] . We shall outline the principal idea of Jagers' model with reference to the binary splitting of cells. The generating function of progeny numbers having the form

$$h(s)=ps^2+1-p$$

mirrors a situation in which each cell at the end of the mitotic cycle faces two possible but mutually exclusive alternatives: either a successful division producing exactly two descendants or else death. The probability of the former is p and of the latter is 1-p. Let us now assume that the moments of cell division and death do not necessarily coincide. In other words, there are two random variables: X_1 - the time until the death of a cell and X_2 - the time until its successful division. Starting from the cell entry into the mitotic cycle, the variables X_1 and X_2 have probability distributions $L_1(x)$ and $L_2(x)$, respectively. There is the extreme form of mutual relationship between these variables since the observed life time of a cell X with the probability 1-p is the variable X_1, and with the probability p it is the variable X_2. Hence, the distribution of X is a simple two-component mixture of the distributions of X_1 and X_2, i.e.

$$L(x) = (1-p)L_1(x) + pL_2(x) . \tag{24}$$

As shown by Jagers, in the case of a generalized supercritical ($p > 0.5$) non-lattice process

$$\frac{M_1(a,t)}{M_1(t)} \sim K(a) , \quad t\to\infty ,$$

where

$$K(a) = \frac{\int_0^a e^{-\alpha t}[1-L(t)]dt}{\int_0^\infty e^{-\alpha t}[1-L(t)]dt} .$$

Taking into account (24) and defining the Malthusian parameter α as the root of the characteristic equation

$$2p \int_0^\infty e^{-\alpha t} dL_2(t) = 1 , \tag{25}$$

we have the following expression for limiting age distribution

$$K(a) = \frac{2\alpha \int_0^a e^{-\alpha t}[1-(1-p)L_1(t)-pL_2(t)]dt}{1-2(1-p) \int_0^\infty e^{-\alpha t} dL_1(t)} . \tag{26}$$

Integrating by parts, taking into consideration (25) and the condition $L_2(0)=0$, it can be easily demonstrated that

$$\int_0^\infty e^{-\alpha t}[1-L_2(t)]dt = \frac{2p-1}{2p\alpha} .$$

From this equality and (26) it is apparent that in the absence of cell death ($p=1$) the function $K(a)$ becomes the limiting distribution $B(a)$.

Besides, it can be said with assurance [6] that if $p > 0.5$ and $L_2(t)$ is a non-lattice distribution the ratio $N(a,t)/N(t)$, referred to as actual age distribution, as $t \to \infty$ converges almost surely to $K(a)$ on a set of non-degenerating trajectories (non-extinction set) of the process $N(t)$.

Using the scheme, Jagers obtained a maximum likelihood estimator of the probability p on the interval $[0,t]$. This estimator may be presented as

$$\hat{p}_t = I_t(2I_t+N_0-N_t), \tag{27}$$

where I_t is the number of mitoses on the interval $[0,t]$, N_0 is the initial number of cells in the population, and N_t is the total number of cells by the moment t. Estimator (27) may be employed both in the case of $L_1(x)=L_2(x)$ which corresponds to the Bellman-Harris process, and in the event cell death occurs before completion of the mitotic cycle. The case when the life time of a cell before death X_1 is long (stochastically longer than X_2) presents considerably greater difficulties from the viewpoint of estimating the variable p and requires extra parametric information. Jagers has shown that for the supercritical non-lattice Bellman-Harris process with $t \to \infty$ the estimator $\hat{p}_t$ almost surely converges to the limit $\hat{p}_\infty$ on a non-extinction set. The number $\hat{p}_\infty$ appears in the form

$$\hat{p}_\infty = \frac{1}{1+2(1-p)\int_0^\infty e^{-\alpha t} dL_1(t)}$$

making it possible to estimate the asymptotic bias of $\hat{p}_t$. Analysis of experimental findings obtained in vitro for certain lines of normal and tumor mammalian cells performed by Jagers [5] has shown that while the p values for the normal cells are within 0.8-0.9 those for the tumor counterparts, as a rule, exceed 0.9.

1.5. The Multitype Age-Dependent Branching Processes

We consider a population which consists of n types of particles (cells). The lifetime of a type i particle is a random variable with distribution $G_i(t)$, $i=1,2,\ldots,n$. To define the particle production of a n-type process, we need n generating functions, each in n variables. The k-th generating function, h_k , will determine the distribution of the number of offspring of various types to be produced by a type k particle. Thus we let

$$h_k(s_1,\ldots,s_n) = \sum_{j_1,\ldots,j_n \geq 0} p_k(j_1,\ldots,j_n) s_1^{j_1} \ldots s_n^{j_n} \qquad (28)$$

$0 \leq s_i \leq 1$, $i=1,2,\ldots,n$,

where $p_k(j_1,\ldots,j_n)$ is the probability that a type k parent produces j_1 particles of type 1, j_2 of type $2,\ldots,j_n$ of type

n.

Further we adopt the following conventions:

1). Points in n-dimensional Euclidean space $\mathbb{R}^n$ are denoted by heavy type: $\mathbf{x}=(x_1,\ldots,x_n)$;

2). $\mathbf{0}=(0,\ldots,0)$, $\mathbf{1}=(1,\ldots,1)$, $\mathbf{s}=(s_1,\ldots,s_n)$;

3). $\mathbf{e}_k=(0,\ldots,0,1,0,\ldots,0)$, with 1 in the k th component;

4). $\mathbf{N}_n$ is n-dimensional lattice space, i.e. the set of all points of $\mathbf{R}^n$ with integer coordinates;

5). $\mathbf{N}_n^+ = \{x \in N_n: x_i \geq 0,\ i=1,\ldots,n\}$;

6). We use the product notation

$$x^y = \prod_{i=1}^{n} x_i^{y_i} .$$

Then for the particle production probability and generation functions we write

$$p(j) = (p_1(j),\ldots,p_n(j)) \tag{29}$$

and

$$h(s) = (h_1(s),\ldots,h_n(s)). \tag{30}$$

Using (29) and (30) we may now rewrite (28) in the form

$$h(s) = \sum_{j \in N_n^+} p(j)s^j \ , \ 0 \leq s \leq 1 \ . \tag{31}$$

Let $Z(t)=(Z_1(t),\ldots,Z_n(t))$ denote the number of particles of the various types existing at time t, and let

$$F_i(t,s) = \sum_{j \in N_n^+} \mathbb{P}\{Z(t)=j\,|\,Z(0)=e_i\}s^j \ ,$$

$$\mathbf{F}(t,s)=(F_1(t,s),\ldots,F_n(t,s)) \ . \tag{32}$$

Then as for the one-dimensional age-dependent process one can show [1,12] that

$$\begin{cases} F_k(t,s)=s_k[(1-G_k(t)] + \displaystyle\int_0^t h_k(\mathbf{F}(t-x,s))dG_k(x) \ , \\ k = 1,2,\ldots,n \end{cases} \tag{33}$$

Indeed, a decomposition of the sample space Ω in accordance with the life-length and number of offspring of the initial particle suggests the relation:

$$\mathbb{P}\{Z(t)=j \mid Z(0)=e_k\}=$$

$$\sum \mathbb{P}\left\{\begin{array}{l}\text{the initial particle of a type } k \text{ dies at time} \\ x \text{ and produces } i=(i_1,\ldots,i_n) \text{ offspring; and} \\ \text{in the remaining time } t-x \text{ these } i \text{ parti-} \\ \text{cles give rise to a total of } j=(j_1,\ldots,j_n) \\ \text{offspring}\end{array}\right\},$$

the sum being taken over $0 \le x \le t$, $i \in N_n^+$. Thus

$$\mathbb{P}\{Z(t)=j_k \mid Z(0)=e_k\} = [1-G_k(t)]\,\delta_{e_k j} +$$

$$+\int_0^t \sum_{i \in N_n^+} p_k(i) \quad \mathbb{P}^{*i_1}\{Z(t-x)= j \mid Z(0)=e_{i_1}\} * \tag{34}$$

$$* \ldots * \mathbb{P}^{*i_n}\{Z(t-x)=j \mid Z(0)=e_{i_n}\}dG_k(x) \ ,$$

where $\mathbb{P}^{*i_k}$ is the i_k-fold convolution of $\mathbb{P}$, and $\delta_{e_k j}$ is the Kronecker delta. The first term on the right side takes care of the case when the initial particle lives longer than time t. Multiplying (34) by s^j and summing over $j \in N_n^+$ we obtain the system (33) of basic equations.

The existence and uniqueness theory of the system (33) can be developed along the lines of the one-dimensional case, i.e. if $G_k(0+)=0$, $k=1,\ldots,n$, and $h_k(s)$ are probability generating functions then (33) has a solution $\mathbf{F}(t,s)=(F_1(t,s),\ldots F_n(t,s))$, where $F_k(t,s)$ are generating functions for each t, and which is the unique bounded solution.

Let $\mathsf{M}=\{m_{ij}\}$ be the particle production mean matrix associated with $h(s)$, i.e.

$$m_{ij} = \frac{\partial}{\partial s_j} h_i(s)\bigg|_{s=1} , \ i,j=1,2,\ldots,n \ . \tag{35}$$

If $\mathsf{M} < \infty$ (which means that $m_{ij}<\infty$ for $i,j=1,2,\ldots,n$) and $G_k(0+)=0$, $k=1,\ldots,n$, then $\mathbb{P}\{Z(t) < \infty\}=1$ for each $t\ge 0$, i.e. the process is regular.

Let $\mathcal{A}(t)=\{A_{ij}(t)\}$ be the matrix of means at time t, i.e.

$$A_{ij}(t) = \mathbb{E}\{Z_j(t) \mid Z(0) = e_i\} = \frac{\partial}{\partial s_j} F_i(t,s)\Big|_{s=1}, \qquad (36)$$

$i,j = 1,2,\ldots,n.$

If $\|\mathcal{M}\| = \max\{ |m_{ij}|;\ i,j=1,2,\ldots,n\} < \infty$, then $\|\mathcal{A}(t)\|$ is bounded on finite intervals, and $\mathcal{A}(t)$ satisfies the matrix equation

$$\mathcal{A}^T(t) = D[\mathbb{1}-\Gamma(t)] + \int_0^t \mathcal{A}^T(t-x)\mathcal{M}^T d[\Gamma(x)], \qquad (37)$$

where $D[\mathbb{1}-\Gamma(t)]$ is the diagonal matrix with $1-G_i(t)$ in the i-th place, and $d[\Gamma(x)]$ is the diagonal matrix with $dG_k(x)$ in the k-th entry. In fact, $\mathcal{A}(t)$ is the unique solution bounded on finite intervals.

There is an analog to the concept of Malthusian parameter introduced in the one-dimensional case. Let $\hat{\mathcal{M}}(\alpha)$ be the matrix whose (i,j) entry is $m_{ij} \int_0^{\infty} e^{-\alpha t} dG_i(t)$. The Malthusian parameter is that number (unique, if it exists) such that the maximal eigenvalue of $\hat{\mathcal{M}}(\alpha)$ is 1. Denote by ρ the maximal eigenvalue of $\mathcal{M}$. In the supercritical case $(\rho > 1)$, the Malthusian parameter α always exists and $\alpha > 0$. In the subcritical case $(\rho < 1)$ the Malthusian parameter α may not exist but if it does, $\alpha < 0$. Finally, in the critical case $(\rho=1)$ we have $\alpha = 0$.

It can be shown that there exist a constant matrix $\mathcal{C}$ and a matrix $\mathcal{B}(t)$ such that

$$\mathcal{A}(t) = \mathcal{C}e^{\alpha t}+\mathcal{B}(t), \qquad (38)$$

where $\|\mathcal{B}(t)\| = O(e^{\beta t})$ for some $\beta < \alpha$.

Note that when $G_k(t)$, $k=1,2,\ldots,n$, is the unit step function

$$G_k(t) = \begin{cases} 0 & \text{for } t < 1 \\ 1 & \text{for } t \geq 1 \end{cases}, \qquad (39)$$

then (33) reduces to a functional iteration formula for $h(s)$ of type (4), i.e. in this case the process $Z(t)$ is a multi-type Galton-Watson branching process.

If

$$G_k(t) = \begin{cases} 0 & \text{for } t < 0, \\ 1-e^{-\lambda_k t} & \text{for } t \geq 0, k=1,2,\ldots,n, \end{cases} \tag{40}$$

then (33), on differentiation, yields the Kolmogorov equations, i.e. the process $Z(t)$ is a multi-type Markov branching process. Excepting the cases (39) and (40), the age-dependent branching processes are not Markovian (but their limit behaviour is similar to the Markov case).

REFERENCES

1. Assmusen, S. A. and Hering H. Branching Processes, Birkhauser, Boston, 1983.
2. Athreya, K.B. and Ney, P.E. Branching processes, Springer-Verlag,Berlin,1972.
3. Crump, K.S. and Mode, C.J. An age-dependent branching process with correlations among sister cells,J.Appl. Prob., 6, 205-210,1969.
4. Harris, T.E. The theory of branching processes, Springer-Verlag,Berlin,1963.
5. Jagers, P. The composition of branching populations: A mathematical result and its application to determine the incidence of death in cell proliferation, Math. Biosci., 8, 227-238,1970.
6. Jagers,P. Branching processes with biological applications, John Willey and Sons, N.Y. - Sidney - Toronto, 1975.
7. Kimmel, M. General theory of cell cycle dynamics based on branching processes in varying environment, In: Biomathematics and Cell Kinetics, Elsevier/North-Holland Biomedical Press, Amsterdam, New York, Oxford,357-373,1981.
8. Macken, C.A. and Perelson, A.S. Branching processes applied to cell surface aggregation phenomena, Lecture Notes in Biomathematics,Springer-Verlag, Berlin, Heidelberg,New York, 1985.
9. Mode, C.J. Multitype branching processes, American Elsevier, N.Y.,1971.
10. Nooney, G.C. Age distributions in stochastically dividing populations, J.Theor. Biol., 20, 314-320, 1968.
11. Nedelman,J., Downs,H. and Pharr,P. Inference for an age-dependent, multitype branching-process model of mast cells, J.Math.Biol.,25,203-226,1987.
12. Sevastyanov, B.A. Branching processes, Nauka,Moscow, 1971 (In Russian).

II. INDUCED CELL PROLIFERATION KINETICS WITHIN THE FRAMEWORK OF A BRANCHING PROCESS MODEL

2.1. Introduction

Mathematical description of initial stages in the growth of a cell population may be constructed in the following manner. Let us assume that at the moment $t=0$ there is in the population a certain, generally speaking, random number of cells m. At the moment $t=0$ the population is acted upon by the proliferative stimulus and all the cells appear to be at the zero age in relation to the mitotic cycle which they afterwards pass through independently of one another. On completion of the mitotic cycle each cell, independently of other cells, generates the random number ν of descendants which are immediately involved into the next cycle of division. When it is considered that all the cells have the same distribution of the mitotic cycle duration X and the same distribution of the number of daughter cells ν, the variables X and ν being independent, it is then reasonable to turn to the model of the Bellman-Harris process defined by means of the distribution function for mitotic cycle duration

$$G(t) = \mathbb{P}\{X \le t\} \tag{1}$$

and the generating function for the number ν of total cell progeny

$$h(s)=\mathbb{E}\{s^{\nu}\}= \sum_{k=0}^{\infty} \mathbb{P}\{\nu=k\}s^k= \sum_{k=0}^{\infty} P_k s^k, \quad |s|\le 1. \tag{2}$$

Following the lead of Sevastyanov [18], in the sequel we shall call such a branching process the (G,h)-process. Thus, every cell of the initial population (existing at the moment $t=0$) represents the zero-th generation in the corresponding branching age-dependent stochastic process; by the moment $t>0$ descendants of ancestor cells belonging to different later generations

$i=1,2,..,n,...$ will have accumulated in the population.

On the assumption of independent evolution of individual cells it is possible to study the dynamics of the population under consideration as the superposition of m standard (G,h)-processes. With this formalization of induced cell proliferation a number of practically interesting results can be obtained. Specifically, it is possible to investigate representation of different generations of ancestor cells in the total population at any instant t. This problem usually arises in studying biochemical events of the mitotic cycle by means of synchronized cell populations. The chapter is based on our publications [26-28].

2.2. The Subsequent Generations of Cells Induced to Proliferate

Let us first consider the case when the process starts with one cell of zero age. We shall define the random process (field) $\mu_n(t)$ as the number of cells that have participated in the formation of the n-th generation at the moment $t\geq 0$. In other words, if we "freeze" the process in the n-th generation, prohibiting transition of cells into the (n+1)-th generation, $\mu_n(t)$ will then represent the size of the n-th generation in the "frozen" process. Let us try to find how characteristics of the process $\mu_n(t)$ vary with time, provided $\mu_0(0)=1$. To this end we shall derive a recurrent formula which enables calculation of the generating function of the process $\mu_n(t)$, i.e.

$$\Phi_n(s;t) = \mathbb{E}\{s^{\mu_n(t)} \,|\mu_0(0)=1\}, \quad t\geq 0 \ , \quad |s| \leq 1. \qquad (3)$$

Let us first find the following conditional mathematical expectation

$$\mathbb{E}\{ s^{\mu_{n+1}(t)} \,|\, \omega \} ,$$

where $\omega = (X,\nu)$ is the evolution of the initial cell.

Then

$$\Phi_{n+1}(s;t) = \int_\Omega \mathbb{E}\{s^{\mu_{n+1}(t)} \,|\, \omega \}dP(\omega) ,$$

where Ω - is the space of all evolutions ω, $P(\omega)$ is the probability measure on that space defined by the functions (1) and (2).

With $X>t$ it is obvious that $\mu_{n+1}(t)=0$, $n\geq 0$, hence

$$\mathbb{E}\{s^{\mu_{n+1}(t)} \mid X > t\} = 1 . \tag{4}$$

If $X\leq t$ and $\nu \geq 1$ then holds the relation

$$\mu_{n+1}(t,\omega) = \sum_{i=1}^{\nu} \mu_n(t-X,\omega_i)$$

where $\mu_n(t-X,\omega_i)$ are independent random variables whose distributions coincide with that of $\mu_n(t-X,\omega)$. Thus, with $X\leq t$

$$\mathbb{E}\{s^{\mu_{n+1}(t)} \mid \nu, X\leq t\} = \Phi_n^{\nu}(s;t-X) . \tag{5}$$

Averaging conditional expectations (4) and (5) over the distribution (X,ν), we obtain the required recurrent formula

$$\begin{aligned}
&\Phi_{n+1}(s;t) = 1-G(t)+ \int_{0-}^{t+} h[\Phi_n(s;t-u)]dG(u) ,\\
&\Phi_0(s;t)\equiv s,\\
&n = 0,1,2,\ldots ; \ |s| \leq 1; \ t\geq 0.
\end{aligned} \tag{6}$$

By way of illustration we shall calculate by formula (6) probabilities for $\mu_n(t)$ in the first and second generations of the initial cell in the case of binary splitting, i.e. $h(s)=s^2$. By successively substituting the functions $\Phi_0(s;t)$ and $\Phi_1(s;t)$ in (6) we obtain for this case

$$\Phi_1(s;t)=1+(s^2-1)G(t) ;$$

$$\Phi_2(s;t)=1+2G^{*2}(t)+b(t)-2[G^{*2}(t)-b(t)]s^2+b(t)s^4,$$

where

$$G^{*2}(t)=[G * G](t)=\int_0^t G(t-u)dG(u) ; \ b(t)=\int_0^t G^2(t-u)dG(u) .$$

Therefore eventually we have

$$\mathbb{P}\{\mu_1(t)=0\} = 1-G(t)$$

$$\mathbb{P}\{\mu_1(t)=2\} = G(t)$$

$$\mathbb{P}\{\mu_2(t)=0\} = 1-2G^{*2}(t)+b(t)$$

$$\mathbb{P}\{\mu_2(t)=2\} = 2[G^{*2}(t)-b(t)]$$

$$\mathbb{P}\{\mu_2(t)=4\} = b(t) ,$$

where G^{*2} and $b(t)$ were defined in the foregoing. In a similar way probabilities for the variable $\mu_n(t)$ can also be calculated for later generations. Let us now pass to calculating the fist two moments of the random field $\mu_n(t)$.

On the natural assumption of $\mathbb{P}\{\nu\leq\infty\}=1$ the regularity of the process $\mu_n(t)$ is guaranteed at any finite n . Thus,

$$\mathbb{P}\{\mu_n(t) < \infty\}=1 , \quad t\geq 0 ,$$

hence

$$\lim_{s\to\infty} \Phi_n(s;t) = 1 .$$

Let us introduce the following designations

$$M_n(t) = \mathbb{E}\{\mu_n(t)\} = \left.\frac{\partial\Phi_n(s;t)}{\partial s}\right|_{s=1} ,$$

$$V_n(t) = \mathbb{E}\{\mu_n(t)[\mu_n(t)-1]\} = \left.\frac{\partial^2\Phi_n(s;t)}{\partial s^2}\right|_{s=1} .$$

Differentiating (6) once and twice, respectively, at the point $s=1$ and assuming the values $\eta=h'(1)$ and $\zeta=h''(1)$ to be finite, we come to the following expressions

$$M_{n+1}(t) = \eta \int_0^t M_n(t-u)dG(u), \quad M_0(t)=1 ; \tag{7}$$

$$V_{n+1}(t)=\zeta \int_0^t M_n^2(t-u)dG(u)+\eta \int_0^t V_n(t-u)dG(u) ; \quad V_0(t)=0. \tag{8}$$

It may be inferred from (7) that the expectation of the

function $\mu_n(t)$ i.e. $M_n(t)=\mathbb{E}\{\mu_n(t)\}$ has the form

$$M_n(t) = \eta^n G^{*n}(t) , \tag{9}$$

where $G^{*n}(t)$ defines the n-fold convolution of the distribution function $G(t)$, with $G^{*1}(t)=G(t)$ and $G^{*0}(t)=1$.

The same recurrent relation (8) holds for the second initial moment

$$R_n(t)=\mathbb{E}\{\mu_n^2(t)\} = V_n(t)+M_n(t) ,$$

but with altered initial condition:

$$R_{n+1}(t) = \zeta \int_0^t M_n^2(t-u)dG(u) + \eta\int_0^t R_n(t-u)dG(u) , \tag{10}$$

$$R_0(t)=1.$$

The variance of $\mu_n(t)$ may be obtained from the formula

$$D_n(t)=\mathbb{D}\{\mu_n(t)\}=R_n(t)-M_n^2(t) . \tag{11}$$

It is easily seen from (9) that

$$\lim_{t\to\infty} M_n(t)=\eta^n. \tag{12}$$

As regards the limiting value of the variance $D_n(t)$ the following result can be readily established.

If the parameters η and $\sigma^2=\zeta+\eta-\eta^2$ are finite, then with $t\to\infty$ the variance $D_n(t)$ tends to the Galton-Watson process variance, i.e. to the value

$$\mathcal{D}_n =\begin{cases} \dfrac{\sigma^2\eta^n(\eta^n-1)}{\eta^2-\eta} & , \eta\neq 1 \\ n\sigma^2 & , \eta=1 . \end{cases} \tag{13}$$

Indeed, coming to the limit in the formula resulting from successive substitutions of $R_0(t)$, $R_1(t)$,...,$R_n(t)$ in formula (10), by simple manipulations we get

$$\lim_{t\to\infty} R_n(t)=\zeta\eta^{n-1}(1+\eta+\eta^2+...+\eta^{n-1})+\eta^n$$

whence

$$\lim_{t\to\infty} R_n(t) = \begin{cases} \zeta\eta^{n-1} \left(\dfrac{\eta^n-1}{\eta-1} \right), & \eta\neq 1, \\ n\zeta+1, & \eta=1. \end{cases} \tag{14}$$

Substituting $\zeta = \sigma^2 - \eta + \eta^2$ in (14) and using formula (11) we obtain

$$\lim_{t\to\infty} D_n(t) = \begin{cases} \dfrac{\sigma^2\eta^n(\eta^n-1)}{\eta^2-\eta}, & \eta\neq 1, \\ n\sigma^2, & \eta=1, \end{cases}$$

i.e. the value $\mathcal{D}_n$.

If the number of descendants ν is equal to the constant number k with probability one, i.e. $h(s)=s^k$, then $\sigma^2=0$ and from (13) it follows that

$$\lim_{t\to\infty} D_n(t) = 0.$$

Results (12) and (13) can also be obtained heuristically, seeing that with $t\to\infty$ the process $\mu_n(t)$ should tend to the Galton-Watson process [11].

By the same way as was used in deriving formula (6) recurrent relation can be obtained for the generating function for the joint distribution of the values $\mu_n(t)$ and $\mu_{n+1}(t)$

$$\Phi_{n,n+1}(s_1,s_2;t)=\mathbb{E}\{s_1^{\mu_n(t)}\ s_2^{\mu_{n+1}(t)}\ |\mu_0(0)=1\},\ |s_1|\le 1,\ |s_2|\le 1.$$

This relation has the form

$$\Phi_{n,n+1}(s_1,s_2;t)=1-G(t)+\int_0^t h[\Phi_{n-1,n}(s_1,s_2;t-u)]dG(u),$$

$$\Phi_{0,1}(s_1,s_2;t)=s_1[1-G(t)+h(s_2)G(t)],$$

$$n=1,2,\ldots;\ |s_1|\le 1;\ |s_2|\le 1;\ t\ge 0.$$

Differentiating $\Phi_{n,n+1}(s_1,s_2;t)$ once with respect to s_1 and once with respect to s_2 and assuming $s_1 = s_2 = 1$, we obtain a recurrent formula for the product moment of the second order

$L_{n,n+1}(t) = \mathbb{E}\{\mu_n(t)\mu_{n+1}(t)\}$:

$$L_{n,n+1}(t)=\eta\int_0^t L_{n-1,n}(t-u)dG(u)+\zeta\int_0^t M_{n-1}(t-u)M_n(t-u)dG(u) ,$$

$$L_{0,1}(t)=\eta\, G(t) .$$

Hence, in view of formula (9), for the covariation

$$C_{n,n+1}(t)=COV(\mu_n(t)\mu_{n+1}(t))=L_{n,n+1}(t)-M_n(t)M_{n+1}(t)$$

the following relationship may be written

$$C_{n,n+1}(t)=\eta\int_0^t C_{n-1,n}(t-u)dG(u) +$$

$$\eta^{2n-1}(\eta+\zeta)\int_0^t G^{*(n-1)}(t-u)G^{*n}(t-u)dG(u) -$$

$$\eta^{2n+1}G^{*n}(t)G^{*(n+1)}(t) .$$

Now the results obtained may be generalized for the case when the process starts with m cells of the zero age, m being non-random. In that case the generating function of $\mu_n(t)$ will have the form

$$\mathbb{E}\{s^{\mu_n(t)} | \mu_0(0)=m\} = \Phi_n^m(s;t) . \tag{15}$$

Differentiating (15) at the point $s=1$ and introducing the designation for mathematical expectation

$$M_n^{<m>}(t) = \mathbb{E}\{ \mu_n(t) | \mu_0(0)=m\} ,$$

we obtain

$$M_n^{<m>}(t) = mM_n(t) . \tag{16}$$

Differentiating (15) twice with respect to s, assuming $s=1$ and using formula (11), we have a similar result for variance

$$D_n^{<m>}(t) = mD_n(t) . \tag{17}$$

Thus, with the superposition of m of branching processes the mean value $M_n(t)$ and the variance $D_n(t)$ are multiplied by the number m , while the variation coefficient of the function $\mu_n(t)$ decreases $\sqrt{m}$ times.

Further generalization of the model may be achieved, assuming that the initial number of cells m is a random variable with the probability distribution

$$q_k = \mathbb{P}\{m=k\} \ , \ k=1,2,\ldots \ ,$$

and the generating function

$$Q(s) = \sum_{k=0}^{\infty} q_k s^k \ .$$

Let us introduce designations for the first and second initial moments of the variable m

$$\alpha_1 = \sum_k k q_k \ ; \ \alpha_2 = \sum_k k^2 q_k \ .$$

The variance of m will be denoted with the symbol $D\{m\}$. Let us now find the generating function $\Gamma_n(s;t)$ of the process $\mu_n^{<m>}(t)$ equal to the sum of a random number of the independent variables $\mu_n(t)$, i.e.

$$\mu_n^{<m>}(t) = \sum_{i=1}^{m} \mu_{n,i}(t) \ , \tag{18}$$

where the second subscript on $\mu(t)$ indicates the number assigned to every process which is a part of sum (18) and the variable m is assumed to be independent of the items $\mu_{n,i}(t)$. On the basis of (18) we can write

$$\Gamma_n(s;t) = Q[\Phi_n(s;t)] \ .$$

Now it is possible to take the expected value $M_n^{<m>}(t)$ and the variance $D_n^{<m>}(t)$ of the function $\mu_n^{<m>}(t)$. Indeed,

$$M_n^{<m>}(t) = \left. \frac{\partial \Gamma_n(s;t)}{\partial s} \right|_{s=1} = Q'(1)\Phi_n'(1,t) \ ,$$

whence

$$M_n^{<m>}(t) = \alpha_1 M_n(t) \ . \tag{19}$$

Similarly, for the variance we obtain

$$D_n^{<m>}(t) = \alpha_1 D_n(t) + (\alpha_2 - \alpha_1^2) M_n^2(t)$$

or

$$D_n^{<m>}(t) = \alpha_1 D_n(t) + D\{m\} M_n^2(t) . \qquad (20)$$

Substituting $\alpha_1=m$ and $D\{m\}=0$ in (19) and (20) we arrive at formulas (16) and (17) for a fixed (deterministic) initial population size.

Distribution of cells induced to division among different generations may also be studied by means of another sequence of random functions.

Let $Z_n(t)$ be the number of cells in $0,1,\ldots,(n-1)$th generation existing at the moment t, where $n=1,2,\ldots$. Assume that $Z_0(t)\equiv 0$. The generating functions for $Z_{n-1}(t)$ and $Z_n(t)$ are linked by the following recurrent relation [11]

$$\Psi_{n+1}(s;t) = s[1-G(t)] + \int_0^t h[\Psi_n(s;t-u)]dG(u)$$

$$\Psi_0(s;t)\equiv 1 \qquad (21)$$

$$n=0,1,2,\ldots; \quad |s|\le 1; \quad t>0.$$

Hence, e.g. for expectation, we obtain

$$A_n(t) = \mathbb{E}\{Z_n(t)\} = \sum_{k=1}^{n} \eta^{k-1}[G^{*(k-1)}(t)-G^{*k}(t)]. \qquad (22)$$

Variations with time of the size of a given n-th generation may be investigated by means of characteristics of the field $Y_n(t) = Z_{n+1}(t)-Z_n(t)$. Specifically, from (22) it follows that the mean size of the n-th generation existing at the moment t is equal to

$$B_n(t) = \mathbb{E}\{Y_n(t)\} = \eta^n[G^{*n}(t)-G^{*(n+1)}(t)] \qquad (23)$$

or, returning to the definition of the process $\mu_n(t)$

$$\mathbb{E}\{Y_n(t)\} = M_n(t) - \frac{1}{\eta} M_{n+1}(t) . \qquad (24)$$

Expression (23) is a good illustration of the conceptual connection between the processes $Y_n(t)$ and $\mu_n(t)$.

Obviously, the variables Z_{n+1} and Z_n cannot be considered as independent and for calculating the variance $D\{Y_n(t)\}$ it is necessary to know the product moment

$$K_{n+1,n}(t) = \mathbb{E}\{Z_{n+1}(t)Z_n(t)\} .$$

The recurrent relationship for the generating function of

the joint distribution of the processes $Z_n(t)$ and $Z_{n+1}(t)$ is of the form

$$\Psi_{n,n+1}(s_1,s_2;t)=s_1s_2[1-G(t)]+\int_0^t h[\Psi_{n-1,n}(s_1,s_2;t-u)]dG(u) ,$$

$$\Psi_{0,1}(s_1,s_2;t)=s_2 \qquad (25)$$

$n=1,2,...;\ |s_1|\le 1;\ |s_2|\le 1;\ t\ge o$.

Hence, differentiating with respect to s_1 and s_2 at the point $s_1=s_2=1$ we have the recurrent formula for the function $K_{n+1,n}(t)$:

$$K_{n+1,n}(t)=1-G(t)+\eta\int_0^t K_{n,n-1}(t-u)dG(u) +$$

$$\zeta\int_0^t A_n(t-u)A_{n-1}(t-u)dG(u),\ n\ge 1 ,$$

where

$$K_{1,0}(t) = \mathbb{E}\{Z_1(t)Z_0(t)\} \equiv 0 \quad \text{and} \quad A_0(t) \equiv 0 .$$

The variance $\Theta_n(t)$ of the size of the n-th generation existing at the moment t is calculated on the basis of the formula

$$\Theta_n(t)=\mathbb{D}\{Y_n(t)\}=\mathbb{D}\{Z_{n+1}(t)\}+\mathbb{D}\{Z_n(t)\}+2[A_{n+1}(t)A_n(t)-K_{n+1,n}(t)] .$$

Thus, we have expressed characteristics (the mean and variance) of the process $Y_n(t)$ in terms of the characteristics of the process $Z_n(t)$ .However, following the same basic scheme of reasoning which has led to formulas (6), (21) and (25), a relation may be deduced directly for the generating function $W_n(s;t)$ of the random field $Y_n(t)$

$$W_{n+1}(s;t)=1-G(t)+\int_0^t h[W_n(s;t-u)]dG(u) ,$$

$$W_0(s;t)=G(t)+s[1-G(t)] ;$$

$n=0,1,2,...;\ |s|\le 1\ ;\ t\ge 0$.

Hence, in the usual fashion, the recurrent formula for the variance of $Y_n(t)$ is obtained

$$\Theta_{n+1}(t)=\eta\int_0^t \Theta_n(t-u)dG(u)-\eta\int_0^t B_n(t-u)dG(u)+(\eta+\zeta)\int_0^t B_n^2(t-u)dG(u) .$$

If the process starts with the random number m of cells which then undergo evolution independently of one another, the expectation and variance of the functions $Z_n(t)$ and $Y_n(t)$ are transformed as in (19) and (20) and the product moment $\mathbb{E}\{Z_{n+1}Z_n\}$ in that case is of the form

$$K^{<m>}_{n+1,n}(t)=\alpha_1 K_{n+1,n}(t)+(\alpha_2-\alpha_1)A_{n+1}(t)A_n(t) .$$

Conceptually, the closest to the foregoing approaches is the work by Nedelman et.al. [16] dealing with the construction and use of a stochastic model of mast-cell clones growth for analysis of experimental data. A particularly attractive feature of that work is a harmonius combination of experimental and mathematical tools responsible for obtaining important information on numerical parameter values for the biological subject. The experimental design in that work was as follows. Individual mast cells, referred to as initiator cells, were placed by micromanipulation procedure into mapped locations in culture dishes. At each observation the total number of cells present at each mapped location was counted. Two sets of data were satisfactorily fitted by a model of age-dependent branching process with three types (initiator, proliferative and resting) of cells. The first one was a sample of observations of the time until division for the initiator cell (it was assumed that this time follows Γ-distribution), and the second was the average colony size as a function of time. The authors derived no equations for probabilistic generating functions, confining themselves to description of mean growth trajectory, which is quite justifiable in view of the practical aims of the work. We believe that supplementation of the results of such experiments with radioautographic data and their subsequent analysis based on models discussed in this chapter would provide a more penetrating insight into the temporal organization of cell colony development.

Kharlamov [10] studied the size of generations of one type of particles $Y_n(t)$ in a Markov homogeneous branching process with continuous time and proposed the following formula for the expectation of the number of particles of the n-th generation at the moment t

$$\mathbb{E}\{Y_n(t)\}=\frac{1}{n!}(a\eta t)^n e^{-at}\ ;\ n=0,1,2,\ldots\ . \qquad (26)$$

where a is the parameter of exponential life length distribution of particles.

It goes without saying that this formula is a particular case of expression (23) which holds for a more general age-dependent process. In fact, in the Markov case expression (23) takes the form

$$\mathbb{E}\{Y_n(t)\}=\frac{\eta^n a^n}{(n-1)!}\int_0^t x^{n-1}e^{-ax}dx-\frac{\eta^n a^{n+1}}{n!}\int_0^t x^n e^{-ax}dx\ .$$

Integrating by parts in the second integral, we at once obtain formula (26). Besides, in this case

$$M(t)=\sum_{n=0}^{\infty}\mathbb{E}\{Y_n(t)\}=e^{a(\eta-1)t}\ ,$$

hence, the distribution $\mathbb{E}\{Y_n(t)\}/M(t)$, $n=0,1,2,\ldots$, is Poisson with the parameter $a\eta t$. Kharlamov also examined the random distribution $Y_n(t)/Y(t)$, where $Y(t)=\sum_{n=0}^{\infty}Y_n(t)>0$, and its asymptotic behaviour as $t\to\infty$. He showed that, with appropriate normalization, $Y_n(t)/Y(t)$ converges in probability to normal distribution. More general results were obtained by Samuels [17] within the (G,h)-process model. In particular he considered two random functions: (1) the proportion of the population belonging to the n-th generation at time t , and (2) the proportion of the n-th generation ever born by time t. In the supercritical case these functions were found to have, asymptotically for large t, a normal form. For the critical and subcritical cases similar results hold with the random variables replaced by their expectations.

Good and Smith [8] used formula (26) derived by Kharlamov for studying the growth of cell clones originating from one cell. The authors modified the formula somewhat, assuming that at the end of the mitotic cycle a cell of the n-th generation either divides and is replaced by two cells of the (n+1)-th generation with probability p_n , or loses the ability for reproducing with probability $1-p_n$. Besides, exponential distribution of the form

$F(t)=1-e^{-t}$ was used, i.e. the mean mitotic cycle duration for each generation was assumed to be equal to one. Under such assumptions the mean number of cells of the n-th generation existing at the moment t is calculated by the formula

$$\mathbb{E}\{Y_n(t)\}=\frac{(2t)^n}{n!}\,e^{-t}\,P^{(n-1)}\ ,$$

where $P^{(n)}=\prod_{i=0}^{n} p_i$.

This simple model shows a good agreement with experimental evidence obtained for human diploid fibroblast clones [8] . It predicts, however, a somewhat greater variability of clone life-span than that observed by experiment. The latter fact induced Good [7] to turn to a modified version of the model based on a dislocated (biased) exponential distribution for the mitotic cycle duration. Dependence of probability of losing the ability to divide on the number of generation may be described, for instance, by simple formula [19]

$$p_i=\frac{i}{i+\gamma}$$

where the parameter γ is estimated from experimental data.

Application of the homogeneous Markovian model shows that variations in clone life-time may in principle be accounted for by variations in mitotic cycle duration and by changes in the reproductive ability of cells depending on the generation number [8] . However, there is ample evidence suggesting that mean generation time (mean duration of the mitotic cycle) may also increase with the aging of a cell population [1,9,14]. Formula (23) in principle allows this fact to be taken into account as well as the loss of reproductive ability and cell death before division, although the problem of statistical estimating the model's unknown parameters becomes in that case more indeterminate. Thus, a more profound mathematical description of the growth of cell clones calls for corresponding development of experimental methods for their investigation. In reference [5] a cell population is assumed to be organized in a hierarchy of decreasing proliferative potential and increasing degree of differentiation. The authors considered 3 classes of cells: stem cells, transitional cells and end cells. Using some elements of the theory of multi-type Galton- Watson branching processes the model with three types of cells was formalized and a method was proposed for the estimation of the probability of self-renewal of stem cells from the experimental distribution of clonal unit sizes.

Description of the growth of cell clones is a promising field for applications of modelling and statistics of stochastic processes. In this connection one cannot but wonder at the scarcity of such studies in the literature. Present-day concepts of the regular patterns of cell clone growth are outlined in a recent monograph by Macken and Perelson [13].

2.3. Age Distributions in Successive Generations

Let us consider two point stochastic processes: $N_k(a,t)$ is the number of k-th generation cells whose age is below or equal to a, and $N(a,t)$ is the total number of cells aged not more than a existing at the moment t. It is clear that these two processes are connected by the relation

$$N(a,t) = \sum_{k=0}^{\infty} N_k(a,t).$$

The cell kinetic indicators $N_k(a,t)$ and $N(a,t)$ are given a full probabilistic description, using general methods of the theory of branching age-dependent stochastic processes.

Let us define the generating function of the random field $N_k(a,t)$

$$Q_k(s;a,t)=\mathbb{E}\{s^{N_k(a,t)} \mid N_0(0,0)=1\},$$
$$Q_0(s;a,0)=s;$$
$$|s|\le 1;$$

and show that it satisfies the following recurrent relationship

$$Q_{k+1}(s;a,t)=1-G(t)+\int_0^t h[Q_k(s;a,t-u)]dG(u) \ ,$$
$$Q_0(s;a,t)=G(t)+[1-G(t)]s^{J(a-t)} \ , \qquad (27)$$
$$k=0,1,2,\ldots \ ; \ |s|\le 1 \ ; \ t\ge 0 \ ;$$

where the function $J(u)$ is defined by the conditions: $J(u)=1$ if $u\ge 0$ and $J(u)=0$ if $u<0$.

Let us calculate the conditional expectation

$$\mathbb{E}\{s^{N_{k+1}(a,t)} \,|\, (X,\nu)\} \ ,$$

where (X,ν) is the evolution of the initial cell.

With $X>t$ we have

$$N_{k+1}(a,t)=0 \ , \ k\ge 0 \ ; \ N_0(a,t)=J(a-t) \ ,$$

hence,

$$\mathbb{E}\{s^{N_{k+1}(a,t)} \,|\, X>t\}=1 \ ,$$
$$\mathbb{E}\{s^{N_0(a,t)} \,|\, X>t\}=s^{J(a-t)} \ . \qquad (28)$$

With $X\le t$

$$N_{k+1}(a,t) = \sum_{i=1}^{\nu} N_k^{(i)}(a,t-X) \ ,$$
$$N_0(a,t)=0 \ ,$$

where $N_k^{(i)}(a,t-X)$ are the mutually independent identically distributed random variables whose distribution coincides with that of $N_k(a,t-X)$. Thus

$$\mathbb{E}\{s^{N_{k+1}(a,t)} \,|\, (X\le t,\nu)\}=Q_k^{\nu}(s;a,t-X) \ ,$$
$$\mathbb{E}\{s^{N_0(a,t)} \,|\, (X\le t,\nu)\}=1 \ . \qquad (29)$$

Averaging conditional expectations (28) and (29) over the distribution (X,ν), we arrive at relations (27).

Differentiating (27) with respect to s at the point $s=1$, we obtain for the age distributions $M_k(a,t)=\mathbb{E}\{N_k(a,t)\}$ the recurrent formula

$$M_{k+1}(a,t)=\eta\int_0^t M_k(a,t-u)dG(u) \quad , \qquad (30)$$
$$M_0(a,t)=J(a-t)[1-G(t)] \ .$$

In analyzing experimental data on the kinetics of synchronized cell populations one usually draws on the second of formulas (30).

It is precisely in the same way that one deduces the integral equation for the generating function of the process N(a,t) that we are familiar with from Chapter I

$$F(s;a,t)=[1-G(t)]s^{J(a-t)}+\int_0^t h(F(s;a,t-u)]dG(u) \ ,$$
$$F(s;a,0)=s \ ;$$
$$|s|\le 1 \ ; \ t\ge 0 \ ;$$

and for the expected number of cells aged below or equal to a

$$M(a,t)=J(a-t)[1-G(t)]+\eta\int_0^t M(a,t-u)dG(u) \ ,$$

where

$$M(a,t)=\mathbb{E}\{N(a,t)\} \ .$$

The formulas can be easily extended to the case of an initial population consisting of a random number m of cells (see 2.2.).

2.4. A Multitype Branching Process Model and Induced Cell Proliferation Kinetics

Processes of preparation for DNA synthesis initiated in cells that have responded to proliferative stimuli develop throughout a portion of the cell cycle called the prereplicative period (PS - period). The mean duration of the period varies substantially depending upon the type of cells and the character of the stimulus applied [6].Pooling data from biochemical studies of systems with induced proliferation (SIP), Baserga [3,4] proposed to distinguish early and late prereplicative periods. Terskikh et al. [22,23] have refined this classification, presenting evidence that in the early prereplicative period events are taking place associated with "transformation" of a resting cell into a proliferating one, whereas the late prereplicative period is the true G_1 phase of the first mitotic cycle following the effect of the stimulus. Thus, the early prereplicative period, or the period of transformation, may be legitimately regarded as a complex of intracellular processes essential for acquiring by a cell the competence to proliferate. In Chapter V of this monograph further proof is presented of the existence of the period of transformation obtained by comparative studies of the kinetics peculiar to the initial and repeat transitions of cells to DNA synthesis in SIP.

Let us return to the description of SIP dynamics by means of the theory of branching age-dependent stochastic processes. We shall now consider the problem of computing probability characteristics of the number of cells that by the moment t have entered the S-phase of the first and second (after stimulation) mitotic cycles. Let $N^c_{S'}(t)$ and $N^c_{S''}(t)$ denote the respective stochastic processes. Clearly, in that case it is expedient to use a multidimensional (multi-type) branching process model. Let us set up a correspondence between the periods PS, S and G_2+M of the first cycle and the cell types T_1, T_2 and T_3 whose life-time distribution functions are equal to $F_1(t)=F_{PS}(t), F_2(t)= F_S(t)$ and $F_3(t)=F_{G_2+M}(t)$, respectively, and between the periods G_1 and S of the second cycle and the cell types T_4 and T_5 with life-time distributions $F_4(t)=F_{G_1}(t)$ and $F_5(t)=F_2(t)= F_S(t)$. The scheme of cell evolution thus defined will be referred to as an open scheme. It will be assumed that durations of individual cell cycle periods are independent. Let us introduce the vector

$$N^i(t)=(N^i_1(t),\dots,N^i_n(t))$$

where the component $N^i_j(t)$ is the number of T_j type cells existing at the moment t, provided there was one T_i type cell in the system at the initial moment t=0. Let us also introduce multidimentional generating functions

$$h(s)=(h_1(s),\dots,h_n(s)) \ ,$$

$$\Psi(s;t)=(\Psi_1(s;t),\dots,\Psi_n(s;t))$$

where

$$h_i(s)=\mathbb{E}\{s^{\nu_i}\}\ ,\ \Psi_i(s;t)=\mathbb{E}\{s^{N^i(t)}\}\ ,\ s=(s_1,\dots,s_n).$$

The generating functions $\Psi(s;t)$ at $|s|\le 1$ and $t\ge 0$ satisfy the set of non-linear integral equations (see [16] and Section 1.5 of this book)

$$\begin{aligned}&\Psi_k(s;t)=s_i[1-F_i(t)]+\int_{0-}^{t+}h_i[\Psi(s;t-u)]dF_i(u)\ ,\\ &\Psi_i(s;0)=s_i\ ;\ i=1,\dots,n\ .\end{aligned}\tag{31}$$

Under the above conditions with five types of cells, in order to describe the number of cells of the type T_5 born before the moment t let $\Psi_5(s;t)=s_5$. The components of vector $h(s)$ will be defined as

$$h_1(s)=s_2;\ h_2(s)=s_3;\ h_3(s)=1-p+p(s_4)^2;\ h_4(s)=s_5\ ,$$

where $(1-p)$ is the probability of cell death in mitosis or of irreversible entry of both daughter cells into the resting state. The system of equations (31) for the open scheme under consideration takes the form

$$\begin{aligned}&\Psi_1(s;t)=s_1[1-F_1(t)]+\int_0^t\Psi_2(s;t-u)dF_1(u)\\ &\Psi_2(s;t)=s_2[1-F_2(t)]+\int_0^t\Psi_3(s;t-u)dF_2(u)\\ &\Psi_3(s;t)=s_3[1-F_3(t)]+(1-p)F_3(t)+p\int_0^t\Psi_4^2(s;t-u)dF_3(u)\\ &\Psi_4(s;t)=s_4[1-F_4(t)]+\int_0^t\Psi_5(s;t-u)dF_4(u)\\ &\Psi_5(s;t)=s_5\ .\end{aligned}\tag{32}$$

Assuming on (32) $s_5=z$ and $s_1=s_2=s_3=s_4=1$ we obtain for generating functions

$$\psi_i(z;t)=\mathbb{E}\{z^{N_5^i(t)}\}=\Psi_i(1,1,1,1,z;t)$$

the following system of equations

$$\psi_1(z;t)=1-F_1(t)+\int_0^t \psi_2(z;t-u)\,dF_1(u)$$

$$\psi_2(z;t)=1-F_2(t)+\int_0^t \psi_3(z;t-u)\,dF_2(u)$$

$$\psi_3(z;t)=1-pF_3(t)+p\int_0^t \psi_4^2(z;t-u)\,dF_3(u) \qquad (32')$$

$$\psi_4(z;t)=1-F_4(t)+\int_0^t \psi_5(z;t-u)\,dF_4(u)$$

$$\psi_5(z;t)=z\ ,$$

from which we can easily find the function

$$\psi_1(z;t) = \mathbb{E}\{z^{N_5^1(t)}\} = \mathbb{E}\{z^{N_{S''}^c(t)}\}$$

we are interested in. The function is of the form

$$\psi_1(z;t)=1-2p(1-z)F_4 * K(t)+p(1-z)^2F_4^2 * K(t)\ , \qquad (33)$$

where

$$K(t)=F_1 * F_2 * F_3(t)=F_{PS+S+G_2+M}(t)\ .$$

Including in the consideration the initial population size m (i.e. turning to the superposition of the random number m of the processes $N_5^1(t)$), on the basis of (33) we obtain by the standard procedure the following equations for expectation and variance

$$\mathbb{E}\{N_{S''}^{c\langle m\rangle}(t)\}=\alpha_1 2pF_{G_1} * K(t)\ ,$$

$$\mathbb{D}\{N_{S''}^{c\langle m\rangle}(t)\}=\alpha_1 2pF_{G_1} * K(t)+[\mathbb{D}\{m\}-\alpha_1][2pF_{G_1} * K(t)]^2 +$$

$$\alpha_1 2pF_{G_1}^2 * K(t)\ .$$

Deduced in exactly the same way are equations for the expectation and variance of the number of cells that by the moment t have entered the S-phase for the first time after the effect of proliferative stimulus

$$\mathbb{E}\{N_{S'}^{c<m>}(t)\}=\alpha_1 F_{PS}(t) \ ,$$
$$\mathbb{D}\{N_{S'}^{c<m>}(t)\}=\alpha_1 F_{PS}(t)+[D\{m\}-\alpha_1]F_{PS}^2(t) \ . \qquad (34)$$

The process $N_{S'}^{c}(t)$ is consistent with experimentally observed number of labelled cells at the time t, provided the cells are in constant contact with ^{3}H-thymidine and with an agent that blocks mitotic division, e.g. colcemid. It is evident from formulas (34) that the characteristics of the number of constantly (continuously) labelled cells in the first mitotic cycle of SIP are determined by the time parameters of the prereplicative period and by the characteristics of the initial size of the population responding to a proliferative stimulus.

The case when initial population cells are asynchronously involved in a mitotic cycle may be easily described by defining the function of probability distribution of moments at which a cell may pass from the resting state to the zeroth age of the mitotic cycle: $F_T(t)$. In other words, $F_T(t)$ is the distribution function of transformation period duration. Assuming that the duration of the G_1 period is independent of the moment cells are stimulated to divide, then formulas (34) take the form

$$\mathbb{E}\{N_{S'}^{c<m>}(t)\}=\alpha_1 F_T * F_{G_1}(t) \ ,$$
$$\mathbb{D}\{N_{S'}^{c<m>}(t)\}=\alpha_1 F_T * F_{G_1}(t)+[D\{m\}-\alpha_1][F_T * F_{G_1}(t)]^2 \ .$$

According to what is commonly referred to as the Smith and Martin model [21] the functions $F_T(t)$ and $F_{G_1}(t)$ are of the form

$$F_T(t)=1-e^{-\lambda t} \ , \ \lambda=\text{const} \ ,$$
$$F_{G_1}(t)=\begin{cases} 1 \ , & t\geq t_0 \ , \\ 0 \ , & t<t_0 \ . \end{cases}$$

Thus, the Smith and Martin model predicts the following character of variations of the functions $\mathbb{E}\{N_{S'}^{c<m>}(t)\}$ and $\mathbb{D}\{N_{S'}^{c<m>}(t)\}$

$$\mathbb{E}\{N_{S'}^{c}(t)\}=\alpha_1[1-e^{-\lambda(t-t_0)}] ;$$

$$\mathbb{D}\{N_{S'}^{c}(t)\}=\alpha_1[1-e^{-\lambda(t-t_0)}]+[\mathbb{D}\{m\}-\alpha_1][1-e^{-\lambda(t-t_0)}]^2 .$$

It would be of interest to have characteristics of the process $N_S^c(t)$, i.e. the number of cells that have entered the S-period by the moment t, without fixing the number of the mitotic cycle of cells induced to proliferate. In the case the system of equations (31) is no longer applicable. Let $\mu_j^i(t)$ be the number of type T_j cells born up to the moment t from the initial type T_i cell $(i,j=1,2,...,n)$. Not including, for convenience, the initial cell in the size of the population, we shall introduce the generating functions of the process $\mu^i(t)=(\mu_1^i(t),...,\mu_n^i(t))$ as

$$\varphi_i(s;t)=\mathbb{E}\{s^{\mu^i(t)}\} ; \varphi_i(0,t)=1 .$$

Then, considering evolution of the initial cell progeny as we did in the preceding section when deducing relation (6), the following system of the equations may be obtained for the functions $\varphi_i(s;t)$

$$\varphi_i(s;t)=1-F_i(t)+\int_0^t h_i[s\varphi(s;t-u)]dF_i(u) , \qquad (35)$$

where the symbol $s\varphi$ denotes $(s_1\varphi_1,...,s_n\varphi_n)$.

To find the generating function of the process $N_S^c(t)$ the scheme of cell evolution should be slightly modified. Let us distinguish in the period PS two periods: the transformation period T and the G_1 period, assigning to them the cell types T_1 and T_2 with the life-time distribution functions $F_1(t)=F_T(t)$ and $F_2(t)=F_{G_1}(t)$. Cells in the S-period will be classified with the type T_3. Assigned to the G_2+M period will be the T_4 type. Cells of the types T_3 and T_4 have life-time distribution functions $F_3(t)=F_S(t)$ and $F_4(t)=F_{G_2+M}(t)$, respectively. The generating function $h(s)$ will be defined as

$$h_1(s)=s_2 ; h_2(s)=s_3 ; h_3(s)=s_4 ; h_4(s)=1-p+ps_2^2 .$$

The cell cycle structure thus defined will be called a closed scheme. The system of equations (35) in that case takes the form

$$\varphi_1(s;t)=1-F_1(t)+s_2\int_0^t \varphi_2(s;t-u)dF_1(u)$$

$$\varphi_2(s;t)=1-F_2(t)+s_3\int_0^t \varphi_3(s;t-u)dF_2(u) \tag{36}$$

$$\varphi_3(s;t)=1-F_3(t)+s_4\int_0^t \varphi_4(s;t-u)dF_3(u)$$

$$\varphi_4(s;t)=1-pF_4(t)+ps_2^2\int_0^t \varphi_2^2(s;t-u)dF_4(u) \ .$$

In the final analysis we are interested in the generating function of the process $\mu_3^1(t)$

$$\beta_1(z;t)=\mathbb{E}\{z^{\mu_3^1(t)}\}=\mathbb{E}\{z^{N_S^c(t)}\} \ ,$$

which may be obtained by simple manipulations from (36), setting $s_3=z, s_1=s_2=s_4=1$. The eventual result is represented in the form of integral equations

$$\beta_1(z;t)=1-F_1(t)+\int_0^t \beta_2(z;t-u)dF_1(u)$$

$$\beta_2(z;t)=1-(1-z)F_2(t)-zpG(t)+zp\int_0^t \beta_2^2(z;t-u)dG(u) \ , \tag{37}$$

where $G(t)=F_2 * F_3 * F_4(t)$ is the function of distribution of mitotic cycle duration, i.e. of the sum of the G_1, S and G_2+M phase durations. Differentiating (37) with respect to z at the point z=1, we obtain the equation for mathematical expectation

$$M_1(t) = F_1 * F_2(t)+2pM_1 * G(t) \ ,$$

which is a renewal equation whose solution is of the form

$$M_1(t) = \sum_{n=0}^{\infty} (2p)^n F_1 * F_2 * G^{*n}(t) \ .$$

Hence, for the expectation of the number of cells that have entered the S-period up to the moment t we have

$$\mathbb{E}\{N_S^{c<m>}(t)\}=\alpha_1\sum_{n=0}^{\infty}(2p)^n F_{PS} * G^{*n}(t) .$$

Similarly derived is the expression for variance

$$\mathbb{D}\{N_S^{c<m>}(t)\}=\alpha_1\sum_{n=1}^{\infty}(2p)^n[2A(t)+A^2(t)] * F_T * G^{*n}(t) +$$

$$\alpha_1\sum_{n=0}^{\infty}(2p)^n F_{PS} * G^{*n}(t) +$$

$$[\mathbb{D}\{m\}-\alpha_1][\sum_{n=0}^{\infty}(2p)^n F_{PS} * G^{*n}]^2 ,$$

where

$$A(t) = \sum_{n=0}^{\infty}(2p)^n F_{G_1} * G^{*n}(t) .$$

Equations (31) have made it possible to calculate probability characteristics of the number of constantly labelled cells in the first mitotic cycle. Solution of such a problem as applied to pulse label conditions, i.e. construction of a generating function of the number of cells in the S-period of the first mitotic cycle at the moment t – $N_{S'}(t)$ – is also based on using the system of equations (31). Fixing the first cycle, we rewrite system (31) as

$$\Psi_1(s;t) =s_1[1-F_1(t)]+\int_0^t \Psi_1(s;t-u)dF_1(u)$$

$$\Psi_2(s;t) =s_2[1-F_2(t)]-F_2(t) .$$

Thus, the generation function of the number of pulse-labelled cells in the first cycle is

$$\mathbb{E}\{z^{N_{S'}(t)}\}=1-F_{PS}(t)+F_{PS+S}(t)+z[F_{PS}(t)-F_{PS+S}(t)] . \qquad (38)$$

Hence, for the expectation of the process $N_{S'}^{<m>}(t)$ follows the expression

$$\mathbb{E}\{N_{S'}^{<m>}(t)\}=\alpha_1[F_{PS}(t)-F_{PS+S}(t)] . \qquad (39)$$

Formula (39) is sometimes used for estimating temporal parameters of mitotic cycle phases in cell systems stimulated to DNA synthesis [12]. The variance of the $N^{<m>}_{S'}(t)$ process computed on the basis of (38) is

$$\mathbb{D}\{N^{<m>}_{S'}(t)\}=\alpha_1[F_{PS}(t)-F_{PS+S}(t)]+[D\{m\}-\alpha_1][F_{PS}(t)-F_{PS+S}(t)]^2. \quad (40)$$

From formulas (39) and (40) immediately follow the following properties of the variance $\mathbb{D}\{N^{<m>}_{S'}(t)\}$:

1. If $D\{m\}>\alpha_1$, then $\mathbb{D}\{N^{<m>}_{S'}(t)\} > \mathbb{E}\{N^{<m>}_{S'}(t)\}$;

2. If $D\{m\}<\alpha_1$, then $\mathbb{D}\{N^{<m>}_{S'}(t)\} < \mathbb{E}\{N^{<m>}_{S'}(t)\}$;

3. If $D\{m\}=\alpha_1$, then $\mathbb{D}\{N^{<m>}_{S'}(t)\} = \mathbb{E}\{N^{<m>}_{S'}(t)\}$.

These properties indicate how the variability of the initial number of cells affects the variability of the number of cells pulse-labelled in the first cycle.

Using the system of equations (32′) and replacing its last equation with

$$\psi_5(z;t)=z[1-F_5(t)]+F_5(t),$$

we obtain an expression for the generating function of the number of cells in the S-phase of the second (following stimulation) mitotic cycle at the moment t

$$\psi_1(z;t)=1-2p(1-z)[F_{G_1} * K(t)-F_{G_1+S} * K(t)] +$$

$$p(1-z)^2[F_{G_1}(t)-F_{G_1+S}(t)]^2 * K(t) .$$

Hence follow the formulas that describe the expectation and variance of the process $N^{<m>}_{S''}(t)$

$$\mathbb{E}\{N^{<m>}_{S''}(t)\}=2p\alpha_1(F_{G_1}-F_{G_1+S}) * K(t)$$

$$\mathbb{D}\{N^{<m>}_{S''}(t)\}=2p\alpha_1(F_{G_1}-F_{G_1+S}) * K(t)+2p\alpha_1(F_{G_1}-F_{G_1+S})^2 * K(t) +$$

$$[D\{m\}-\alpha_1][2p(F_{G_1}-F_{G_1+S}) * K(t)]^2 .$$

The function K(t) was defined when deriving formula (33). The expression for $\mathbb{E}\{N_{S''}^{<m>}(t)\}$ may be rewritten in the following form more convenient for comparing with expression (39)

$$\mathbb{E}\{N_{S''}^{<m>}(t)\}=2p\alpha_1(F_{PS}-F_{PS+S}) * G(t) .$$

Let us now consider the problem of probabilistic description of the number of cells in the S-period of <u>any</u> mitotic cycle at the moment t , i.e. the number of cells pulse-labelled with ^{3}H-thymidine at the moment t. Denoting the stochastic process under study by $N_S(t)$, we shall turn again to system of equations (31), using the closed evolution scheme for four types of cells described in the foregoing. In this case we have the system of equations for generating functions

$$\Psi_1(s;t)=s_1[1-F_1(t)]+\int_0^t \Psi_2(s;t-u)dF_1(u)$$

$$\Psi_2(s;t)=s_2[1-F_2(t)]+\int_0^t \Psi_3(s;t-u)dF_2(u)$$

$$\Psi_3(s;t)=s_3[1-F_3(t)]+\int_0^t \Psi_4(s;t-u)dF_3(u)$$

$$\Psi_4(s;t)=s_4[1-F_4(t)]+(1-p)F_4(t)+p\int_0^t \Psi_2^2(s;t-u)dF_4(u).$$

Expressions for the expectation and variance of the process $N_S^{<m>}(t)$ are

$$\mathbb{E}\{N_S^{<m>}(t)\}=\alpha_1\sum_{n=0}^{\infty}(2p)^n(F_{PS}-F_{PS+S}) * G^{*n}(t) ,$$

$$\mathbb{D}\{N_S^{<m>}(t)\}=\alpha_1\sum_{n=1}^{\infty}(2p)^n B^2 * F_T * G^{*n}(t)+\alpha_1\sum_{n=0}^{\infty}(2p)^n(F_{PS}-F_{PS+S}) * G^{*n}(t) +$$

$$[\mathbb{D}\{m\}-\alpha_1][\sum_{n=0}^{\infty}(2p)^n(F_{PS}-F_{PS+S}) * G^{*n}(t)]^2 ,$$

where

$$B(t) = \sum_{n=0}^{\infty}(2p)^n(F_{G_1}-F_{G_1+S}) * G^{*n}(t) .$$

The principles underlying the theory of multidimensional branching processes were successfully applied to cell kinetics by Mode [15]. Specifically, Mode has shown that if a mitotic cycle is broken into k successive independent phases, in the supercritical case and assuming $F_i'(t)=f_i(t) \in L_2(0,\infty)$, $i=1,\ldots,k$, there exists with probability one a finite limit on the nonextinction set

$$\lim_{t\to\infty} I_i(t) = y_i,$$

where

$$I_i(t) = \frac{N_i(t)}{N_\Sigma(t)}, \quad N_\Sigma(t) = \sum_{i=1}^{k} N_i(t) .$$

For the case when $f_i(t)$ is the densities of gamma-distribution with a scale parameter identical for all i, Mode gives concrete expressions for the y_i values.

2.5. Grain Count Distribution and Branching Stochastic Processes

2.5.1. Introduction

The microscopic investigation of a radioautograph allows one to count not only the fractions of cells in period S and M of the mitotic cycle at the given instant but the number of grains in each labelled single cell as well. As the interval between the moment of pulse labelling and the moment of tissue specimen fixation is increased, the distribution of this number undergoes certain changes due to the processes of proliferation and desintegration of cells. The information of such changes can be useful in the quantitative analysis of cell kinetics, particularly in estimating the mitotic cycle temporal parameters. It is obvious that the theoretical derivation of the time-varying distribution of the number of grains (marks) per cell would be very useful for applied analysis of autoradiographical observations.

Setting aside such practical questions as metabolic stability and reutilization of a label as well as the correction of experimental data for threshold errors, this problem will be considered in the following idealized formulation proposed earlier by Williams [24].

Assume that the evolution of a cell population is governed by the process of binary splitting and death of cells only. At the initial moment $t=0$ each cell of a given population is instantly labelled with a random number of discrete marks, and the distribution of this number is assumed to be Poisson with parameter Θ :

$$\Pi_j(0) = \frac{\Theta^j}{j!} e^{-\Theta}, \quad j = 0,1,2,\ldots \quad (41)$$

Then we accept the following natural assumption: when a cell divides in two, each of the marks from the mother cell is distributed independently and with probability $\frac{1}{2}$ among daughter cells.

Introduce a stochastic process $Z_j(t)$ to represent a number of cells bearing exactly j marks at the moment t, and define the distribution $\Pi_j(t)$ as the ratio of the mathematical expectations of the processes $Z_j(t)$ and $\Sigma_{j=0}^{\infty} Z_j(t)$. That is,

$$\Pi_j(t) = \frac{N_j(t)}{N(t)}, \quad (42)$$

where

$$N_j(t) = \mathbb{E}\{z_j(t)\}, \quad N(t) = \mathbb{E}\left\{\sum_{j=0}^{\infty} Z_j(t)\right\}.$$

Williams [24] considered the cell population dynamics within the model of a linear nonhomogeneous birth-death process (Markov case) and derived the following expression for the distribution (2):

$$\Pi_j(t) = \frac{\Theta^j}{j!} \exp[-2\Lambda(t)] \sum_{n=0}^{\infty} (-1)^n \frac{\Theta^n}{n!} \exp[2^{1-n-j}\Lambda(t)], \quad (43)$$

where $\Lambda(t)=\int_0^t \lambda(u)du$, with $\lambda(t)$ a rate of cellular multiplication. In the case of a homogeneous birth-death process, $\Lambda(t)=\lambda t$ and (43) is evidently simplified to

$$\Pi_j(t) = \frac{\Theta^j}{j!} \exp[-2\lambda t] \sum_{n=0}^{\infty} (-1)^n \frac{\Theta^n}{n!} \exp[2^{1-n-j}\lambda t]. \quad (44)$$

The most important deduction from the results (43) and (44) is that the distribution $\Pi_j(t)$ does not contain any information about the value of the death rate μ. Under certain experimental conditions this parameter can be found from the equation for the expected total number of cells in a population, which under the conditions of a homogeneous birth-death process has the form

$$N(t) = N(0) \exp [(\lambda-\mu)t] .$$

The aim of this chapter is to generalize the result (44) by means of the formalization of cell population kinetics within the bounds of the time-homogeneous but age-dependent (i.e. non-Markov case) branching process model.

2.5.2. The Distribution $\Pi_j(t)$ for the Bellman-Harris Process

The structure of the Bellman-Harris age-dependent branching process is exhaustively defined by means of an arbitrary distribution function $G(t)$ for the duration T of the mitotic cycle

$$G(t) = \mathbb{P}\{T \le t\} , \quad t \in [0,\infty) , \tag{45}$$

and a probability generating function $h(s)$ for the number ν of cell total progeny

$$h(s) = \mathbb{E}\{s^{\nu}\} = \sum_{k=0}^{\infty} \mathbb{P}\{\nu=k\} s^k , \quad |s| \le 1. \tag{46}$$

Certainly, the usual independence hypotheses (see Chapter I) have to be accepted. We assume that $G(0)=0$ and the generating function (46) is of the form

$$h(s) = ps^2 + 1 - p , \tag{47}$$

where p is a probability of the successful binary splitting and $1-p$ is a probability of the reproductive death of a cell. In this case the branching process $Z(t)$, representing the total number of cells in a population at the moment t, is regular and $\mathbb{E}\{Z(t)\}<\infty$ (see [11, Chapter V, Theorem 13.1]).

Consider the situation when at the moment of impulse labelling ($t=0$) the whole population of cells is synchronized at the starting point (all cells are of zero age) of the mitotic cycle.

Then to describe the evolution of the population of marks, the Bellman-Harris vector branching process with the account set of particle types

$$Z(t) = (Z_0(t), Z_1(t), \dots, Z_n(t), \dots) \tag{48}$$

can be introduced. This process is defined by the distribution function (45) and multidimensional generating functions of the progeny of the nth-type particle

$$h_n(s) = 1 - p + p \sum_{i=0}^{n} \binom{n}{i} 2^{-n} s_i s_{n-i} , \tag{49}$$

where

$$s = (s_0, s_1, \dots, s_k, \dots) , \; n=0,1,\dots \quad .$$

Introduce generating vector function $F(t;s)$ of the process $Z(t)$ with components

$$F_k(t;s) = \mathbb{E}\{s^{Z(t)} | Z(0)=e_k\} , \; k=0,1,2,\dots \quad ,$$

where $s^{Z(t)} = \prod_{k=0}^{\infty} s_k^{Z_k(t)}$, and e_k is infinite vector with the kth component equal to 1 and all other components equal to 0. The condition $Z(0)=e_k$ implies that the process starts from one zero-aged particle of type k. The generating functions $F_k(t;s)$ satisfy the following system of integral equations (see formula (33), Chapter I)

$$F_k(t;s) = s_k[1-G(t)] + \int_0^t h_k[F(t-u;s)]dG(u) , \tag{50}$$
$$|s| \le 1 , \quad k=0,1,\dots \; .$$

Denote the expectation of the number of particles of the jth type originating from one particle of the kth type by

$$A_{kj}(t) = \mathbb{E}\{Z_j(t) | Z_k(0)=1\} = \left. \frac{\partial F_k(t;s)}{\partial s_j} \right|_{s=1} .$$

Substituting (49) in (50), we have

$$F_k(t;s) = s_k[1-G(t)]+(1-p)G(t) + p2^{-k}\sum_{i=0}^{k}\binom{k}{i}\int_0^t F_i(t-u);s)F_{k-i}(t-u;s)dG(u) \quad ,$$

whence the following system of equations for the expectations $A_{kj}(t)$ is readily obtained:

$$A_{kj}(t)=\delta_{kj}[1-G(t)]+p2^{1-k}\sum_{i=0}^{k}\binom{k}{i}\int_0^t A_{ij}(t-u)dG(u) \quad , \quad k \geq j \, , \tag{51}$$

$$A_{kj}(t)=0, \; k<j \, ,$$

There as usual $\delta_{kj}=0$ when $j\neq k$ and $\delta_{kj}=1$ when $j=k$.

Consider the situation when at the moment $t=0$ there exists a certain random number of particles with mathematical expectation equal to M. Then the expectation of the number of jth-type particles at the moment $t\geq 0$, on the strength of assumption (41) is equal to

$$N_j(t,\Theta)=\mathbb{E}\{Z_j(t)\}=Me^{-\Theta}\sum_{k=j}^{\infty}\frac{\Theta^k}{k!}A_{kj}(t) \; . \tag{52}$$

By substituting of the expression (52) into the system (51) we obtain

$$N_j(t,\Theta)=M\frac{e^{-\Theta}\Theta^j}{j!}[1-G(t)] + Mpe^{-\Theta}\sum_{k=i}^{\infty}\frac{2^{1-k}\Theta^k}{k!}\sum_{i=0}^{k}\binom{k}{i}\int_0^t A_{ij}(t-u)dG(u) \; , \tag{53}$$

where the summation over i in effect begins at j in view of the zero terms in (51). After a change in the summation order in (53) we have

$$N_j(t,\Theta)=M\frac{e^{-\Theta}\Theta^j}{j!}[1-G(t)] + M2pe^{-\Theta/2}\int_0^t\sum_{i=j}^{\infty}\frac{(\Theta/2)^i}{i!}A_{ij}(t-u)dG(u) \; . \tag{54}$$

Keeping in mind the definition (52), the equation (54) may be rewritten in a more convenient form

$$N_j(t,\Theta)=M\,\frac{e^{-\Theta}\Theta^j}{j!}\,[1-G(t)]+2p\int_0^t N_j(t-u,\Theta/2)\,dG(u)\ . \qquad (55)$$

Note that the relationship (55) may be treated as a recurrent one in regard to the parameter Θ. Hence iterating the equation (55) n times, we obtain

$$N_j(t,0)=M\,\frac{\Theta^j}{j!}\sum_{k=0}^{n}(2p)^k\exp\left[-\frac{\Theta}{2^k}\right]2^{-kj}\,\bar{G} * G_k(t)\ + \qquad (56)$$

$$(2p)^{n+1}\int_0^t N_j\left(t-u,\frac{\Theta}{2^{n+1}}\right)dG_{n+1}(u)\ ,$$

where $\bar{G}=1-G$, and symbol $*$ designates onefold convolution, i.e.

$$\bar{G} * G_k(t)=\int_0^t \bar{G}(t-\tau)\,dG_k(\tau)\ ,$$

and $G_k(t)$ is defined as follows:

$$G_0(t)\equiv\begin{cases}1, & t\ge 0,\\ 0, & t<0,\end{cases}$$

$$G_{k+1}(t)=\int_0^t G_k(t-u)\,dG(u)=G_k * G(t)\ ,\ k=0,1,2,\ldots\ ,\ \text{for } t\ge 0.$$

With the help of (52) the second term in the expression (56) is evaluated as follows:

$$I(n)=(2p)^{n+1}\int_0^t N_j\left(t-u,\frac{\Theta}{2^{n+1}}\right)dG_{n+1}(u)\ =$$

$$(2p)^{n+1}\left[N_j\left(\cdot,\frac{\Theta}{2^{n+1}}\right) * G_{n+1}\right](t)\ =$$

$$(2p)^{n+1}M\exp\left[-\frac{\Theta}{2^{n+1}}\right]\sum_{k=j}^{\infty}\frac{\Theta^k}{k!\,2^{k(n+1)}}\left[A_{kj} * G_{n+1}\right](t)\ \le$$

$$p^{n+1}M\exp\left[-\frac{\Theta}{2^{n+1}}\right]\sum_{k=j}^{\infty}\frac{\Theta^k}{k!}\left[A_{kj} * G_{n+1}\right](t)\ .$$

Now we have inequality

$$I(n) \le e^{\Theta} p^{n+1} \left[N_j(\cdot,\Theta) * G_{n+1}\right](t) \le e^{\Theta} p^{n+1} \left[N * G_{n+1}\right](t) \ . \qquad (57)$$

In consequence of the regularity of the process $Z(t)$ and the definition (45) the inequality (57) permits us to assert that $I(n) \to 0$ as $n \to \infty$ for any fixed value of $t \ge 0$. Then, letting n go to infinity, from (56) we find

$$N_j(t,\Theta) = \frac{M\Theta^j}{j!} \sum_{k=0}^{\infty} (2p)^k \exp\left[-\frac{\Theta}{2^k}\right] 2^{-kj} \left[\bar{G} * G_k\right](t). \qquad (58)$$

The expected total number of cells in a population, $N(t) = \sum_{j=0}^{\infty} N_j(t,\Theta)$, is described by the formula

$$N(t) = M \sum_{k=0}^{\infty} (2p)^k \left[\bar{G} * G_k\right](t) \ . \qquad (59)$$

Coming back to definition (42) of the distribution $\Pi_j(t)$ and using (58) and (59), we derive finally the expression

$$\Pi_j(t) = \frac{\Theta^j}{j!} \ \frac{\sum_{k=0}^{\infty} (2^{1-j}p)^k \exp\left\{-\frac{\Theta}{2^k}\right\} \left[\bar{G} * G_k\right](t)}{\sum_{k=0}^{\infty} (2p)^k \left[\bar{G} * G_k\right](t)} \ . \qquad (60)$$

Another form of the formula for this distribution is

$$\Pi_j(t) = \frac{e^{-\Theta}\,\Theta^j}{j!} \ \times$$

$$\left\{ \frac{1 + \sum_{k=1}^{\infty} (2^{1-j}p)^k \left(1 - \frac{2^j \exp\left[-\Theta/2^k\right]}{2p}\right) \exp\left[-\frac{\Theta}{2^k}\right] G_k(t)}{1 + \left(1 - \frac{1}{2p}\right) \sum_{k=1}^{\infty} (2p)^k G_k(t)} \right\},$$

which can be obtained by using the relation

$$\bar{G} * G_k(t) = G_k(t) - G_{k+1}(t) \ .$$

It is well known (see [11]) that the regular Bellman-Harris process in the case of $G(t)=1-e^{-at}$ degenerates into a Markov branching process. In turn by allowing

$$a = \lambda + \mu \ , \ p = \frac{\lambda}{\lambda + \mu} \ , \tag{61}$$

we revert to the model of linear homogeneous birth-death process (see [11,Chapter V,§7 and Example 13.1]). Now using (61) and the fact that

$$\bar{G} * G_k(t) = G_k(t) - G_{k+1}(t) = \frac{(at)^k}{k!} e^{-at} \ ,$$

it is not difficult to transform the expression (60) into Williams's result (44).

2.5.3. The Moments of the Distribution $\Pi_j(t)$

On the basis of expression (60) we can now calculate the generating function

$$\psi(t;s) = \sum_{j=0}^{\infty} \Pi_j(t)s^j \ , \quad |s| \le 1 \ ,$$

which has the following explicit form:

$$\psi(t;s) = \frac{\sum_{k=0}^{\infty} (2p)^k \left[\bar{G} * G_k \right](t) \exp\left\{- \frac{\Theta(1-s)}{2^k}\right\}}{\sum_{k=0}^{\infty} (2p)^k \left[\bar{G} * G_k \right](t)} \ . \tag{62}$$

By differentiating (62) n times with respect to s at the point s=1 , we obtain the expression for the nth factorial moment of the distribution (60),

$$\alpha_n(t) = \psi^{(n)}(t;1) = \frac{\Theta^n \sum_{k=0}^{\infty} (2^{1-n}.p)^k \ [\bar{G} * G_k](t)}{\sum_{k=0}^{\infty} (2p)^k \ [\bar{G} * G_k](t)} \ .$$

Hence for the mathematical expectation of the number of marks at the moment t we get

$$\alpha_1(t) = \frac{\Theta \sum_{k=0}^{\infty} p^k \, [\bar{G} * G_k](t)}{\sum_{k=0}^{\infty} (2p)^k \, [\bar{G} * G_k](t)} ,$$

and the variance can be calculated from the formula

$$D(t) = \alpha_2(t) + \alpha_1(t) - \alpha_1^2(t) .$$

From the asymptotic results established for the expectation of the Bellman-Harris process (see Section 1.4) it follows that for every value of the probability p

$$\alpha_n(t) \to 0 \quad \text{as} \quad t \to \infty .$$

2.5.4. Some Generalizations and Applied Problems

Consider now the new process

$$\tilde{Z}(t) = (\tilde{Z}_0(t), \tilde{Z}_1(t), \ldots, \tilde{Z}_n(t), \ldots) , \qquad (63)$$

which differ from the process (48) with respect to characteristics of the zero-generation cells only. Namely, we assume that for the zero generation of cells the distribution function of the mitotic cycle duration is $K(t)=P\{T_0 \le t\}$, $K(0)=0$, instead of $G(t)$, and the probability of successful division is b instead of p. So the probability of reproductive death of a cell at the end of the first mitotic cycle is equal to $1-b$.

Using the well-known methods of branching-process theory (see Section 1.5) it is not difficult to show that the probability generating functions of the process $\tilde{Z}(t)$,

$$\Phi_k(t,s) = \mathbb{E}\{s^{\tilde{Z}(t)} \,|\, \tilde{Z}(0) = e_k\} , \quad k=0,1,2,\ldots,$$

satisfy the following relations:

$$\Phi_k(t,s) = s_k \bar{K}(t) + \int_0^t g_k(F(t-u,s))\,dK(u) , \quad k=0,1,2,\ldots, \qquad (64)$$

where

$$g_k(s) = 1-b+b\sum_{i=0}^{k}\binom{k}{i} 2^{-i} s_i s_{k-i} ,$$

$$\bar{K}(t)=1-K(t) ,$$

and the probability generating functions $F_k(t,s)$ satisfy equations (50). Then from (64) it follows that for the mathematical expectations

$$B_{kj}(t)=\mathbb{E}\{\tilde{Z}_j(t)\,|\,\tilde{Z}_k(0)=1\} = \left.\frac{\partial\Phi_k(t,s)}{\partial s_j}\right|_{s=1}$$

we have

$$B_{kj}(t)=\delta_{kj}(t)\bar{K}(t)+b2^{1-k}\sum_{i=0}^{k}\binom{k}{i}\int_0^t A_{ij}(t-u)dK(u) , \; k \geq j , \tag{65}$$

$$B_{kj}(t)=0, \; k < j ,$$

where the functions $A_{ij}(t)$ satisfy the equations (51).

Now using (65),(51), and (52), the following transformations can be made:

$$\tilde{N}_j(t,\Theta)=\mathbb{E}\{\tilde{Z}_j(t)\}=Me^{-\Theta}\sum_{k=j}^{\infty}\frac{\Theta^k}{k!}B_{kj}(t) =$$

$$Me^{-\Theta}\left[\frac{\Theta^j}{j!}\bar{K}(t)+2b\sum_{k=j}^{\infty}\frac{\Theta^k 2^{-k}}{k!}\sum_{i=j}^{\infty}\binom{k}{i}\int_0^t A_{ij}(t-\tau)dK(\tau)\right] =$$

$$Me^{-\Theta}\frac{\Theta^j}{j!}\bar{K}(t)+2bMe^{-\Theta}\sum_{i=j}^{\infty}\frac{(\Theta/2)^i}{i!}\int_0^t A_{ij}(t-\tau)dK(\tau) =$$

$$Me^{-\Theta}\frac{\Theta^j}{j!}\bar{K}(t)+2b\int_0^t N_j\left(t-\tau,\frac{\Theta}{2}\right)dK(\tau) .$$

Hence by using expression (58) we obtain for any $j=0,1,2,...,$

$$\tilde{N}_j(t,\Theta)=M\frac{\Theta^j}{j!}\left[e^{-\Theta}\bar{K}(t)+2b\sum_{k=0}^{\infty}(2p)^k 2^{-j(k+1)}e^{-\Theta/2^{k+1}}W_k(t)\right] , \tag{66}$$

where $W_k(t)=(\bar{G} * G_k * K)(t)$. For the expected total number of cells in a population the following expression holds:

$$\tilde{N}(t)=\sum_{j=0}^{\infty}\tilde{N}_j(t,\Theta)=M\left[\bar{K}(t)+2b\sum_{k=0}^{\infty}(2p)^k W_k(t)\right]. \quad (67)$$

On the analogy of the derivation of the distribution (60) we finally obtain the following form of the distribution of marks for the modified process $\tilde{Z}(t)$:

$$\tilde{\Pi}_j(t,\Theta) = \frac{\tilde{N}_j(t,\Theta)}{\tilde{N}(t)} =$$

$$\frac{\Theta^j}{j!}\ \frac{\bar{K}(t)e^{-\Theta} + 2b\sum_{k=0}^{\infty}(2p)^k 2^{-j(k+1)}e^{-\Theta/2^{k+1}}W_k(t)}{\bar{K}(t) + 2b\sum_{k=0}^{\infty}(2p)^k W_k(t)}. \quad (68)$$

For our purposes it is necessary to know also the distribution $\tilde{\Pi}_j(t,\Theta|u)$ of marks over a proliferating cell population on the condition that all cells of the initial population are of age u at the moment t=0. In this case from (66) and (67) it follows that

$$\tilde{N}_j(t,\Theta|u)=M\,\frac{\Theta^j}{j!}\left[e^{-\Theta}\,\bar{K}(t,u)+2b\sum_{k=0}^{\infty}(2p)^k 2^{-j(k+1)}e^{-\Theta/2^{k+1}}W_k(t,u)\right] = \quad (69)$$

$$M\,\tilde{N}^1_j(t,\Theta|u)\ ,\quad j=0,1,2,\dots\ ,$$

$$\tilde{N}(t|u)=\sum_{j=0}^{\infty}\tilde{N}_j(t,\Theta|u)=M\left[\bar{K}(t,u)+2b\sum_{k=0}^{\infty}(2p)^k W_k(t,u)\right]=M\,\tilde{N}^1(t|u), \quad (70)$$

where

$$K(t,u) = \mathbb{P}\{T_0 \le t+u \mid T_0 > u\} = \frac{K(t+u) - K(u)}{1 - K(u)},$$

$$\bar{K}(t,u) = 1-K(t,u) \ ,$$

$$W_k(t,u) = \int_0^t K(t-\tau,u)\,dU_k(\tau) \ ,$$

$$U_k(t) = (\bar{G} * G_k)(t) = G_k(t) - G_{k+1}(t) \ .$$

Therefore from (69) and (70) we have for any $j=0,1,2,...$

$$\tilde{\Pi}_j(t,\Theta|u) = \frac{\Theta^j}{j!} \frac{\bar{K}(t,u)e^{-\Theta} + 2b\sum_{k=0}^{\infty}(2p)^k 2^{-j(k+1)} e^{-\Theta/2^{k+1}} W_k(t,u)}{\bar{K}(t,u) + 2b\sum_{k=0}^{\infty}(2p)^k W_k(t,u)} . \quad (71)$$

Now we possess all necessary tools for the consideration of some concrete experimental situations from the viewpoint of applications of the proposed approach.

One of the most evident applications is the estimation of distribution of the duration of the mitotic cycle (or its distinct phases) and cell loss factor in different experimental situations. To achieve this aim it is necessary first of all to provide for the correspondence of the model with the strictly specified state of cell population kinetics. In our case this may be done in the most natural manner if we consider either synchronized populations of cells or systems with induced cell proliferation.

Indeed, consider a population of cells which have been synchronized by sampling during mitosis, i.e. by use of the generally accepted method of synchronization of cultured cells. At the time $t=0$ such a population consists exclusively of cells which are starting the first cell cycle. Assume that then during u units of time the reproduction of cells does not occur or can be evaluated as negligible. To justify this assumption one must either choose an appropriate instant of time $t=u$ or apply a certain blocking agent, e.g. colcemide, to prevent mitotic divisions. The later possibility has been taken into account in the model of the process (63) by introducing particular characteristics $K(t)$ and b for cells in the first mitotic cycle in contrast to recycling cells. So at the moment $t=u$ all cells of this population are of age u, but some distribution of cells among different phases of the first cell cycle exists.

Let I_s be the mean proportion of cells traversing the S-phase in the whole population under study at the moment $t=u$, and choose this very moment for labelling of cells with ^{3}H-thymidine. Now we suppose for a moment that a cell is considered as labelled if it bears at least one mark. It is clear that for the mean number $N_0^S(t,\Theta|u)$ of cells bearing no marks at the time $t>u$ the following equality holds:

$$N_0^S(t,\Theta|u) = (1-I_S)\tilde{N}(t|u) + I_S\tilde{N}_0(t,\Theta|u) , \qquad (72)$$

where $\tilde{N}_0(t,\Theta|u)$ and $\tilde{N}(t|u)$ are defined by (69) and (70). Similarly, for the expected number of cells bearing exactly j marks ($j\geq1$) at the moment $t>u$ we have

$$N_j^S(t,\Theta|u) = I_S\tilde{N}_j(t,\Theta|u) . \qquad (73)$$

From (72) and (73) follows the expression for the distribution of marks in such a synchronized (at $t=0$) population of cell s at any moment $t>u$:

$$\pi_0(t,\Theta|u)=1-I_S+I_S \frac{\tilde{N}_0(t,\Theta|u)}{\tilde{N}(t|u)} = 1-I_S+I_S\tilde{\Pi}_0(t,\Theta|u) ,$$

$$\pi_j(t,\Theta|u)=I_S \frac{\tilde{N}_j(t,\Theta|u)}{\tilde{N}(t|u)} = I_S\tilde{\Pi}_j(t,\Theta|u) , \; j\geq1 . \qquad (74)$$

It must be pointed out once more that the derivation of the expressions (74) was based on the fact that all cells (irrespective of their life-cycle phase) of the synchronized population are of the same age at the moment of labelling.

As distinct from the synchronized populations of the above considered kind, system with induced cell proliferation are characterized by the synchronous entry of only a fraction, say γ, of the population of quiescent cells into the mitotic cycle (see

Chapter V). It is easy to see that in order to describe induced proliferative processes the distribution (74) is to be changed in the following way:

$$\pi_0(t,\Theta|u) = 1-I_S \frac{\tilde{N}^1(t|u)-\tilde{N}_0^1(t,\Theta|u)}{1-\gamma+\gamma\tilde{N}^1(t|u)} ,$$

$$\pi_j(t,\Theta|u) = I_S \frac{\tilde{N}_j^1(t,\Theta|u)}{1-\gamma+\gamma\tilde{N}^1(t|u)} \quad \text{for} \quad j\geq 1 , \tag{75}$$

where $\tilde{N}_0^1(t,\Theta|u)$, $\tilde{N}_j^1(t,\Theta|u)$, and $\tilde{N}^1(t|u)$ are defined by (69) and (70). If two distinct synchronous fractions participate in the cell-system proliferative response [25] then with double-labelling technique one obtains radioautographical data which may be considered in the same manner.

The quantity I_S is just the index of impulse labelling of cells with ^{3}H-thymidine, which can be estimated by means of direct radioautographic observations (see Remark I below). The procedure for evaluating the parameter γ in systems induced to proliferate will be discussed in Chapter V. A suitable estimator for the parameter Θ can be obtained by fitting initial distribution (41) to corresponding experimental data; thus the time-varying distribution (75) (or its special case (74)) may be used for estimating the rest of the parameters which are of interest for the study of systems induced to proliferate (or synchronized cell systems), namely, the probability 1-b of cell death at the end of the first mitotic cycle, the probability 1-p of death of recycling cells, and the temporal parameters (mean and variance of the duration) of the first and subsequent mitotic cycles. [29].

Remark 1. As a rule the observed labelling index I_S^{obs} is defined as the ratio of the number of cells bearing a number of marks which exceeds some specified critical value m to the total cell numbers. It is clear that the quantities I_S^{obs} and I_S do not coincide, but the following evident relation between them holds:

$$I_S^{obs} = I_S \sum_{j=m}^{\infty} \frac{e^{-\Theta}\,\Theta^j}{j!} = I_S \left[1 - \sum_{j=0}^{m} \frac{e^{-\Theta}\,\Theta^j}{j!}\right].$$

Therefore the value I_S can be calculated by use of the formula

$$I_S = \frac{I_S^{obs}}{1 - \sum_{j=0}^{m} \frac{e^{-\Theta}\,\Theta^j}{j!}} .$$

Hence for the case considered above (m=0) the correction of the experimental labelling index has to be made as follows:

$$I_S = \frac{I_S^{obs}}{1 - e^{-\Theta}} .$$

Remark 2. As follows from the work [2], more complex distributions such as a compound Poisson may be a better description of the initial distribution of marks than (41). The corresponding generalization of the main results (60),(68),(71), (74),(75) can be achieved by randomizing the parameter Θ in all these expressions. For example

$$\hat{\Pi}_j = \int_0^\infty \Pi_j(t,x)\,dR(x),$$

where $\hat{\Pi}_j(t)$ is the generalized form of the distribution (60), and $R(x)$ is the *a priori* distribution of the parameter Θ. The case of finite mixtures of Poisson distributions [2] also may be treated in similar manner and does not meet any difficulties.

The next example concerning the applications of time-varying distribution of marks is an exponentially growing cell population. If we consider a supercritical process and $\alpha > 0$ is its Malthusian parameter, then for the index I_S of an exponentially growing population the following formula holds:

$$I_S = \frac{\int_0^\infty e^{-\alpha x}\left[F_{G_1}(x) - F_{G_1+S}(x)\right]dx}{\int_0^\infty e^{-\alpha x}\left[1 - G(x)\right]dx},$$

where $F_{G_1}(t)$ and $F_{G_1+S}(t)$ are the distribution functions (assumed to be nonlattice like $G(t)$) for the durations of phases G_1 and G_1+S of the mitotic cycle, respectively. Using the formula (71) with $K(t)\equiv G(t)$, $b=p$, and well-known results on the limiting age distribution (see Section 1.4) it is not difficult to btain the distribution of marks for any $t>0$:

$$\Pi_0(t) = \frac{\int_0^\infty e^{-\alpha u}\left[F_{G_1}(u)-F_{G_1+S}(u)\right]N_0^1(t|u)du}{\int_0^\infty e^{-\alpha u}\left[1-G(u)\right]N^1(t|u)du} + \frac{\int_0^\infty e^{-\alpha u}\left[1-G(u)+F_{G_1+S}(u)-F_{G_1}(u)\right]N^1(t|u)du}{\int_0^\infty e^{-\alpha u}\left[1-G(u)\right]N^1(t|u)du},$$

$$\Pi_j(t) = \frac{\int_0^\infty e^{-\alpha u}\left[F_{G_1}(u)-F_{G_1+S}(u)\right]N_j^1(t,\Theta|u)du}{\int_0^\infty e^{-\alpha u}\left[1-G(u)\right]N^1(t|u)du}, \quad j=1,2,\ldots,$$

$N_j^1(t,\Theta|u)=\tilde{N}_j^1(t,\Theta|u)$, $N^1(t|u)=\tilde{N}^1(t|u)$ if $K(t)\equiv G(t)$, $b=p$.

Remarks 1 and 2 are valid for the steady exponential population growth as well.

2.5.5. Radiobiological Applications

Intracellular injuries induced by irradiation (or chemical cytotoxic agents) may be interpreted as discrete marks attached to a cell. Considering cell population kinetics in damaged tissue from this point of view it is natural to assume that the value of cell division probability p depends on the number of such marks. So probability generating function (46) is to be replaced by

$$h_n(\mathbf{s})=1-p_n+p_n\sum_{i=0}^{n}\binom{n}{i}2^{-n}s_i s_{n-i}, \qquad (76)$$

where $\mathbf{s}=(s_0,s_1,\ldots,s_k,\ldots)$.

Then we specify the function p_n that fits well the

radiobiological experience

$$p_n = pa^n, \tag{77}$$

where $0 < a \le 1$, $0 < p \le 1$, $n=0,1,2,\ldots$.
To describe $\Pi_j(0)$ we shall use Poisson distribution again and our choice is substantiated by the modern concepts ("target theory") of quantitative radiobiology as well.

Instead of equations (51) now we have

$$A_{kj}(t)=\delta_{kj}\bar{G}(t)+p_k 2^{1-k}\sum_{i=0}^{k}\binom{k}{i}\int_0^t A_{ij}(t-u)\,dG(u), \quad k\ge j,$$

$$A_{kj}(t)=0 , \quad k<j,$$

and relationship (55) is replaced by

$$N_j(t,\Theta)=Me^{-\Theta}\,\frac{\Theta^j}{j!}\,\bar{G}(t) + 2pe^{-\Theta(1-a)}\int_0^t N_j(t-u;\frac{a\Theta}{2})\,dG(u) . \tag{78}$$

Iterating (78) n times, we obtain

$$N_j(t,\Theta) = M\,\frac{e^{-\Theta}\,\Theta^j}{j!}\sum_{k=0}^{n}\left(2p\left(\frac{a}{2}\right)^j\right)^k\left[\,\bar{G} * G_k\right](t)\exp\left\{\frac{\Theta\, a\left(1-\frac{a^k}{2^k}\right)}{2\left(1-\frac{a}{2}\right)}\right\} + \tag{79}$$

$$I(n)\ ,$$

where

$$I(n) = (2p)^{n+1}\exp\left\{-\Theta(1-a)\,\frac{1-\left(\frac{a}{2}\right)^{n+1}}{1-\frac{a}{2}}\right\}\int_0^t N_j\left(t-u;\Theta\left(\frac{a}{2}\right)^{n+1}\right)dG_{n+1}(u).$$

It is not difficult to show that $I(n)\to 0$ as $n\to\infty$ for any fixed value of $t\geq 0$. Therefore, letting n go to infinity, from (79) we find

$$N_j(t,\Theta) =$$

$$\frac{Me^{-\Theta}\,\Theta^j}{j!}\sum_{k=0}^{\infty}\left(2p\left(\frac{a}{2}\right)^i\right)^k\left[\,\bar{G} * G_k\right](t)\exp\left\{\frac{\Theta\, a}{2}\;\frac{1-\left(\frac{a}{2}\right)^k}{1-\frac{a}{2}}\right\}, \tag{80}$$

$j=0,1,2,\ldots$.

The expected total number of cells $N(t)$ is described by the formula

$$N(t) = M\exp\left\{-\Theta\,\frac{(1-a)}{\left(1-\frac{a}{2}\right)}\right\}\sum_{k=0}^{\infty}(2p)^k\left[\,\bar{G} * G_k\right](t)\exp\left\{\frac{\Theta(1-a)}{1-\frac{a}{2}}\left(\frac{a}{2}\right)^k\right\}. \tag{81}$$

From (80) and (81) we derive finally the expression

$$\Pi_j(t) = \frac{\Theta^j}{j!} \; \frac{\sum_{k=0}^{\infty} \left(2p\left(\frac{a}{2}\right)^j\right)^k \exp\left\{-\frac{\Theta}{(1-\frac{a}{2})}\left(\frac{a}{2}\right)^{k+1}\right\}\left[\bar{G} * G_k\right](t)}{\sum_{k=0}^{\infty} (2p)^k \exp\left\{\frac{\Theta(1-a)}{(1-\frac{a}{2})}\left(\frac{a}{2}\right)^k\right\}\left[\bar{G} * G_k\right](t)} \; .$$

In the special case $a=1$ this expression degenerates into (60).

It is impossible to observe the number of radiation injuries in a direct experiment. To measure the biological effect of irradiation the clonogenic capacity of cells is commonly used. This experimental indicator may be considered as an estimator for the extinction probability which must depend on the initial distribution of injuries. Extinction probability is defined as follows

$$r(t;\Theta)=\mathbb{P}\{Z(t)=0\}, \quad Z(t)=\sum_{j=0}^{\infty} Z_j(t) \; . \qquad (82)$$

Introduce generating functions

$$\varphi_k(t;x) = \mathbb{E}\{x^{Z(t)} | Z(0)=e_k\} =$$

$$F_k(t;s)\big|_{s_1=s_2\ldots=s_k=x \;;\; s_{k+1}=s_{k+2}=\ldots = 0} \; .$$

The generating functions $F_k(t;s)$ satisfy the following system of equations

$$F_k(t;s)=s_k[1-G(t)]+(1+p_k)G(t) +$$

$$p_k 2^{-k} \sum_{i=0}^{k} \binom{k}{i} \int_0^t F_i(t-u;s) F_{k-i}(t-u;s)\,dG(u).$$

Therefore for the functions $\varphi_k(t;x)$ we have

$$\varphi_k(t;x)=x[1-G(t)]+(1-p_k)G(t) + \tag{83}$$

$$p_k 2^{-k} \sum_{i=0}^{k} \binom{k}{i} \int_0^t \varphi_i(t-u;x)\varphi_{k-i}(t-u;x)dG(u),$$

$k=0,1,2,\ldots$.

Actually it is necessary to derive the equation for the generating function for the process $Z(t)$ under condition that the process starts from one cell bearing a random number of injuries. We define this function by means of the formula

$$R(t;x|\Theta) = \sum_{k=0}^{\infty} \frac{\Theta^k}{k!} e^{-\Theta}\varphi_k(t;x). \tag{84}$$

From (77),(83) and (84) the desired equation may be obtained

$$R(t;x|\Theta)=x[1-G(t)]+G(t)[1-pe^{-\Theta(1-a)}] +$$

$$pe^{-\Theta(1-a)} \int_0^t R^2 (t-u;x|\frac{a\Theta}{2})dG(u) .$$

Coming back to definition (82) and keeping in mind that $r(t;\Theta)=R(t;0|\Theta)$ we get the equation

$$r(t;\Theta)=G(t)[1-pe^{-\Theta(1-a)}] + \tag{85}$$

$$pe^{-\Theta(1-a)} \int_0^t r^2(t-u;\frac{a\Theta}{2})dG(u) .$$

Unfortunately, the last recurrent relationship is nonlinear and can not be treated in the same manner as (78), but it is probable that asymptotic behaviour of the extinction probability for $t\to\infty$ may be studied on the basis of (85). The generalization of (85) to cover the case of random initial number of damaged cells needs no comment.

REFERENCES

1. Absher, P. M., Absher, R. G. and Barnes,W. D. Time-lapse cinemicrophotographic studies of cell division patterns of human diploid fibroblasts (WI-38) during their in vitro lifespan, In:Cell Impairment in Aging and Development,Plenum Press, New York,91-105,1975.
2. Bartlett,M.S. Distributions associated with cell populations, Biometrika,56,391-400,1969.
3. Baserga,R. Biochemistry of the cell cycle: a review, Cell Tiss.Kinet.,1,167-191,1968.
4. Baserga, R. Multiplication and division in mammalian cells, Academic Press, New York, 1976.
5. Ciampi, A., Kates, L., Buick, R., Kriukov, Y. and Till,J.E. Multy-type Galton-Watson process as a model for proliferating human tumour cell populations derived from stem cells: estimation of stem cell self-renewal probabilities in human ovarian carcinomas, Cell Tiss.Kinet.,19,129-140,1986.
6. Gelfant, S. A new concept of tissue and tumour cell proliferation, Cancer Res.,37,3845-3862,1977.
7. Good, P.J. A note on the generation age distribution of cells with a delayed exponential lifetime, Math.Biosci, 24,21-24, 1975.
8. Good,P.J. and Smith,J.R. The age distribution of human diploid fibroblasts,Biophys.J.,14,811-823,1974.
9. Grove,G.L. and Cristofalo, V.J. The "transition probability model" and the regulation of proliferation of human diploid cell cultures during aging, Cell Tiss.Kinet.,9,395-399,1976.
10. Kharlamov, B.P. On the numbers of particle generations in a branching process with non-overlapping generations, Theor.Probab.Appl.,14,44-50,1969(In Russian).
11. Harris, T.E. The theory of branching processes, Springer-Verlag, Berlin,1963.
12. Koschel, K.W.,Hodgson, G.S. and Radley, J.M. Characteristics of the isoprenaline stimulated proliferative response of rat submaxillary gland, Cell Tiss.Kinet.,9,157-165,1976.
13. Macken, C.A. and Perelson, A.S. Stem cell differentiation. Lecture Notes in Biomathematics, Springer-Verlag, Berlin, Heidelberg, New York, 1988.
14. Macieira-Coelho,A., Ponten, J. and Philipson, L. Inhibition of the division cycle in confluent cultures of human fibroblasts in vitro, Exper.Cell Res.,43,20-29,1966.
15. Mode,C.J. Multitype age-dependent branching processes and cell cycle analysis, Math.Biosci.,10,177-190,1971.
16. Nedelman, J., Downs, H. and Pharr, P. Inference for an age-dependent, multitype branching-process model of mast cells, J.Math.Biol., 25, 203-226, 1987.
17. Samuels ,M.L. Distribution of the branching-process population among generations, J.Appl.Prob., 8, 655-667,1971.
18. Sevastyanov,B. A. Branching processes, Nauka, Moscow,1971 (In Russian).
19. Shall,S. and Stein, W.D. A mortalization theory for the control of cell proliferation and for the origin of immortal cell lines, J.Theor.Biol.,76,219-231,1979.
20. Smith,J.R. and Hayflick,L. Variation of lifespan of clones derived from human diploid cell strains, J.Cell Biol.,62,48-53,1974.

21. Smith, J.A. and Martin, L. Do cells cycle?, Proc. Nat. Acad. Sci. USA, 70, 1263-1267, 1973.
22. Terskikh, V.V. and Malenkov, A. G. Variation of ionic composition, RNA and protein synthesis with proliferation induced in a stationary culture of Chinese hamster cells, Cytology, 15, 868-874 (In Russian).
23. Terskikh, V.V., Zisimovskya, A . I. and Abuladze, M.K. Macromolecular syntheses and cell population kinetics in the course of proliferation induction in a stationary culture of Chinese hamster cells, Cytology, 16, 317-321, 1974 (In Russian).
24. Williams, T. The distribution of inanimate marks over a nonhomogeneous birth-death process, Biometrika, 56, 225-227, 1969.
25. Yakovlev. A.Yu., Malinin, A.M., Terskikh V.V. and Makarova G.F. Kinetics of induced cell proliferation at steady-state conditions of cell culture, Cytobiologie, 14, 279-283, 1977.
26. Yakovlev. A.Yu. and Yanev, N.M. The dynamics of induced cell proliferation within the model of a branching stochastic rocess. I. Numbers of cells in successive generations, Cytology, 22, 945-953, 1980 (In Russian).
27. Yanev, N.M. and Yakovlev, A.Yu. The dynamics of induced cell proliferation within the model of a branching stochastic process II. Some characteristics of the cell cycle temporal organization, Cytology, 25, 818-825, 1983 (In Russian).
28. Yanev, N.M. and Yakovlev, A.Yu. On the distribution of marks over a proliferating cell population obeying the Bellman-Harris branching processes, Mathem. Biosci., 75, 159-173, 1985.
29. Yanev, N.M., Balykin, P.V., Goot, R.E., Zorin, A.V., Tanushev, M.S. and Yakovlev, A.Yu. A method for estimation of probability of cell reproductive death, Studia Biophys., 123, 117-124, 1988 (In Russian).

III. SEMISTOCHASTIC MODELS OF CELL POPULATION KINETICS

3.1. Introduction

From the viewpoint of applications it appears important to modify the approaches to modelling cell population dynamics discussed in the preceding chapter so that they could be extended to a wider range of phenomena without substantially complicating the mathematical aspect of the problem.

A cell population is an aggregate of cells distinguished by a definite property [13]. The choice of such a property (or of several properties) depends, as a rule, on the investigator's interests and on the nature of the problem under study. In any event, however, two events are invariably the focus of attention: the acquisition by the cell of the property concerned and the loss by the cell of that property. In this way a one-to-one correspondence is set up between the notions of population and cell cycle phase. For instance, all the cells synthesizing DNA constitute a population corresponding to the S-phase of the mitotic cycle. One may combine the S and G_2 phases, regarding the biochemical processes occurring in either period as a single process, and study the kinetics of the combined population corresponding to the $(S+G_2)$-phase of the cycle. Distinguishing a population by a two-dimensional criterion may be illustrated by considering a population of cells synthesizing DNA for the first time after the effect of the proliferative stimulus (Chapter II).

In distinguishing certain cell populations there is always some arbitrary element at work which depends on the wealth of experimental data on the tissue under study. In effect, theoretical considerations may dictate the need for a special separation of a population even if it is not revealed in a direct experiment. A set of cell populations combined with regard to regulating influences and interrelations between the populations constitutes a cellular system. Making a sharp distinction between the mitotic and the life cycles of cells, it is expedient to introduce the notion of a transitive cell population. In the preceding chapter we have discussed closed populations. One may classify with them a population of stem cells since it is quite reasonable to assume that their transition from a certain phase (dichophase) of the mitotic cycle towards maturation (differentiation) is directed by a mechanism external to a stem cell population which,however, does not prevent daughter cells from beginning again preparations for division under the action of an appropriate stimulus. No population corresponding to a part of the mitotic cycle is a closed population. Such a population may be referred to as transitive,since a cell is bound to enter into that population and then, sooner or later, to leave it. So, a transitive cell population is one that can be regarded as a time-delay dynamic input-output system with a bounded mean time of delay in it. It appears very difficult, indeed, to define that notion in a clearer and more explicit manner because it is nothing but a way of simplifying the cell system structure convenient for modelling purposes. Collected in transitive populations are cells which are maturing and fulfilling specialized tissue functions, provided the mean duration of their residence in such populations is limited. From this standpoint the notion of a closed population appears to be quite conditional, since it may be regarded that every time a cell divides after completing the cycle, two new cells enter the new cycle. Viewed in that light, a closed population is, apparently, a particular case of the transitive population.

One of the problems arising in application of the theory of branching processes in the field of cell kinetics is due to the fact that the assumption that all the viable descendants of a given cell are bound to enter the next division cycle may prove invalid in real cell systems. It will be shown in Chapter V that while some hepatocytes, after completing mitosis during regeneration of the liver, are involved in the next mitotic cycle, some other's leave the cycle. As this takes place, the sizes of populations of cells stimulated to DNA synthesis and of those leaving the mitotic cycle are controlled quantities depending both on the total number of hepatocytes and on the physiological state of the organ. The uncertain fate of daughter cells also complicates considerably mathematical description of the second wave of labelled mitoses curve whenever there is no reason to regard a population as closed with a known generation coefficient. Of course, one may try to construct a refined model of a controlled nonhomogeneous branching process with several types of cells. Such a model, however, would be too cumbersome and complicated for analytical investigation, on the one hand, and would require a priori information generally unobtainable in experiment, on the other. Without denying the theoretical significance of further development of such models, we shall give preference to another approach, viz., devising methods that would require a minimum of simplifying assumptions and whose application to concrete experimental data would yield new information on the temporal organization of the cell cycle. The principal idea of the approach elaborated in the chapters that follow consists in isolated description of individual components of the cell system and subsequent comparative study of their dynamic characteristics. In so doing, where possible, no hypotheses are advanced concerning the nature of interactions between the components (of which at least one is transitive); on the contrary, analysis of empirical evidence is to provide information sufficient to reveal the basic features of such interactions. An example is found in the isolated study of initial and repeated DNA syntheses in systems with induced cell proliferation (Chapter V). Another example is the analysis of diurnal rhythm in cell proliferation processes by means of frequency characteristics (Section 3.3).

Thus, our aim is primarily to develop methods for modelling transitive cell populations and then to employ those methods for processing experimental findings. It is obvious that since the probabilistic structure of the influx of cells may in this case be greatly diversified, it is impossible, in simulating the dynamics of a transitive cell population, to secure such rigorous axiomatic constructions as in the case of a closed population considered within a framework of a branching process model. Discussed in Section 3.2. are semistochastic models of cell kinetics in a steady state. In such models, on the one hand, certain parameters (e.g., cell cycle phase durations) are assumed to be stochastic, while, on the other, only the behaviour of expected values of some principal variables (e.g., number of cells in a population or age distribution) are investigated. In Section 3.4. we modify population steady state equations in such a way as to enable description of transient processes. However, strictly speaking, such equations of unsteady state of cell kinetics hold true only for expectations, their stochastic counterparts requiring in practical applications a very great body of statistical information. Thus, wider applicability of the mathematical apparatus, i.e. its extension to a broader range of cell kinetics phenomena, is achieved in that case at a sacrifice in the completeness of the probabilistic description of the processes under consideration.

3.2. Integral Equations of Steady-State Dynamics of a Transitive Cell Population

Introduce the following functions of time: $K_1(t)$ - the rate of cell entry into a population (cycle phase), $K_2(t)$ - the rate of cell exit from it, $N(t)$ - the total number of cells in a population.

On a continuous set of non-negative random variables A we shall define the stochastic process $N(a,t)$ which will be taken to mean the number of cells whose age A does not exceed the value a at the moment t. The function $N(a,t)$ may be considered as a random point distribution (see [16] for definition and details). Presented in the preceding chapter was the equations for the generating function of the process $N(a,t)$ which can be constructed in the framework of a model of a branching age-dependent stochastic process. Another approach [1] to studying a population's age composition consists in relating to each age value the probability density $w(a,t)$ and the probability that a given cell at the moment t is of an age from the interval $(a,a+\Delta a)$ is equal to $w(a,t)\Delta a+o(\Delta a)$. Thus, age distribution density may be defined as

$$\rho(a,t)=\mathbb{E}\{dN(a,t)\}/da \,, \tag{1}$$

i.e. $\rho(a,t)$ characterizes "distribution of expectations among ages".

The functions $w(a,t)$ and $\rho(a,t)$ are connected by the relation

$$w(a,t) = \frac{\rho(a,t)}{\mathbb{E}\{N(t)\}}, \tag{2}$$

where $N(t)$ is the total size of the population at the moment t. The distribution density $\rho(a,t)$ satisfies von Foerster's equation [13]

$$\frac{\partial\rho(a,t)}{\partial t} + \frac{\partial\rho(a,t)}{\partial a} = -\lambda(a,t)\rho(a,t), \tag{3}$$

where the function $\lambda(a,t)$, usually referred to as failure rate or loss function, may be expressed in terms of the distribution of the variable X which defines the length of a cell's residence in a population.

Let Y denote the random time of zero-aged cell entering a population, then

$$\lambda(a,t) = \lim_{\Delta a \to 0+} \frac{P\{\, a<X\le a+\Delta a \mid X>a,\ Y=t-a\}}{\Delta a} = \frac{f(a \mid Y=t-a)}{1-F(a \mid Y=t-a)}$$

Here $f(x \mid Y=\tau)$ is the conditional density of distribution for the length of a cell's residence in a population given $Y=\tau$, and $F(x \mid Y=\tau)$ is the corresponding conditional distribution function. By changing the variables: $\tau=t-a$, we have

$$\lambda(a,\tau+a) = \frac{f(a \mid Y=\tau)}{1-F(a \mid Y=\tau)}.$$

It then follows from the last expression that

$$1 - F(a|Y=\tau) = \exp [-\int_0^a \lambda(u,\tau+u)du].$$

Therefore we may write

$$1-F(a|Y=t-a) = \exp [- \int_0^a \lambda(u,t-a+u)du] .$$

Equation (3) is solved by changing the variables and reducing it to an ordinary differential equation. For the boundary conditions

$$\rho(o,t) = \mathbb{E}\{K_1(t)\} , \rho(a,0) = \xi(a) ,$$

the solution (see [36] for details)

$$\rho(a,t)=\mathbb{E}\{K_1(t-a)\}\exp [-\int_0^a \lambda(u,t-a+u)du], t>a,$$

(4)

$$\rho(a,t)=\xi(a-t)\exp [-\int_{a-t}^a \lambda(u,t-a+u)du], t<a,$$

where $K_1(t)$ is the rate of entry of new cells into the population and $\xi(a)$ is the initial distribution of cells' age. It is evident from (4) that von Forester's equation makes it possible to describe not only closed but also transitive cell populations, this being due to additional definition of the function $K_1(t)$.

It is clear that the density $\rho(a,t)$ obtained from equation (3) permits calculation of only the mean number of cells in a certain interval of age values. For obtaining higher moments of $N(a,t)$ it is necessary to consider joint probability distributions known as product densities [1]. For instance, the product density of the first order

$$\rho(a_1,a_2;t) = \mathbb{E}\{dN(a_1,t)dN(a_2,t)\}/da_1da_2 , a_1\neq a_2 ,$$

enables investigation of the second moment of $N(a,t)$.

Corresponding to the density function $\rho=\rho(a_1,a_2;t)$ is the continuity equation

$$\frac{\partial\rho}{\partial t} + \frac{\partial\rho}{\partial a_1} + \frac{\partial\rho}{\partial a_2} = -(\lambda_1 + \lambda_2)\rho .$$

Let us put the first of equations (4) into a somewhat different form, assuming that the length of time a cell spends in a population is independent of the instant it entered the population. In that case

$$\rho(a,t) = \mathbb{E}\{K_1(t-a)\}[1-F(a)] . \tag{5}$$

In order to find on the basis of (5) a steady-state distribution of $\rho(a,t)$ the function $\mathbb{E}\{K_1(t)\}$ should be defined under the conditions of steady-state kinetics. Bartlett [2] explored the asymptotic properties of an age distribution described by (3) in a closed population. Confining to $\lambda(a,t)\equiv\lambda(a)$ and defining $\mathbb{E}\{K_1(t)\}$ as the linear functional

$$\mathbb{E}\{K_1(t)\}=\eta\int_0^\infty \lambda(a)\rho(a,t)da,$$

where η is the offspring mean (the mean number of descendants resulting from the division of a cell), then with t going to infinity

$$\rho(a,t)\sim C_0e^{p(t-a)}\exp[-\int_0^a \lambda(u)du], \; C_0>0 . \tag{6}$$

The parameter p is the positive root of the characteristic equation

$$\eta\int_0^\infty e^{-pt}\lambda(t)[-\int_0^t \lambda(u)du]dt=1 . \tag{7}$$

Using functions $f(x)$ and $F(x)$ introduced in the foregoing, it is easy to rewrite expressions (6) and (7) in the following form

$$\rho(a,t) \sim C_0 e^{p(t-a)}[\ 1-F(a)\ ],\ t\to\infty\ , \tag{8}$$

$$\eta\int_0^\infty e^{-pt} f(t)dt=1\ . \tag{9}$$

The same result is obtained in considering a closed population from the standpoint of a branching age-dependent random process (Chapter II).

Using formulas (2) and (5), we can write

$$w(a,t)=\mathbb{E}\{K_1(t-a)\}[1-F(a)]/\mathbb{E}\{N(t)\}\ . \tag{10}$$

In the sequel it will be assumed that the distribution $F(x)$ has continuous density $f(x)$, whereas $w(a,t)$ is continuous in respect to both variables. Taking into consideration the normalizing condition

$$\int_0^\infty w(a,t)da=1,$$

it is easy to obtain from (10) the equation connecting expectations of the processes $N(t)$ and $K_1(t)$ under the conditions of steady-state cell kinetics

$$\mathbb{E}\{N(t)\}=\int_0^\infty \mathbb{E}\{K_1(t-a)\}[1-F(a)]da\ . \tag{11}$$

By changing the variables $t-a=\tau$ equation (11) is reduced to the form

$$\mathbb{E}\{N(t)\}=\int_{-\infty}^t \mathbb{E}\{K_1(\tau)\}[\ 1-F(t-\tau)\ ]d\tau\ . \tag{12}$$

Differentiating (12) with respect to the parameter t and taking into account the relation of balance

$$\dot{N}(t)=K_1(t)-K_2(t)\ ,$$

we may also obtain the equation for expected rates

$$\mathbb{E}\{K_2(t)\}=\int_{-\infty}^{t}\mathbb{E}\{K_1(\tau)\}f(t-\tau)d\tau \ . \qquad (13)$$

In the sequel instead of the expectation symbol $\mathbb{E}\{\cdot\}$ lower case letters will be used:

$$k_1(t)=\mathbb{E}\{K_1(t)\} \ , \ k_2(t)=\mathbb{E}\{K_2(t)\} \ , \ n(t)=\mathbb{E}\{N(t)\} \ .$$

Equation (12) describes a certain linear stationary system with the input "signal" $k_1(t)$ and the output "signal" $n(t)$. Similarly, corresponding to equation (13) is a system with the input $k_1(t)$ and the output $k_2(t)$. The probabilitstic characteristics $[1-F(x)]$ and $f(x)$ in the light of this interpretation have the meaning of weight (impulse transition) functions of the respective systems. Thus, for describing processes of cell kinetics we shall confine ourselves to the class of linear stationary dynamic systems, attempting to exploit as far as possible the potentialities of such a description for analysis of biological experimental evidence.

Let us now consider the problem of the uniqueness of the solution of integral equation (12) introducing the new kernel $\psi(t-\tau)=[1-F(t-\tau)]J(t-\tau)$, where $J(t-\tau)$ is the unit step function, and reducing (12) to the form

$$n(t)=\int_{-\infty}^{\infty}k_1(\tau)\psi(t-\tau)d\tau \ . \qquad (14)$$

In order to substantiate the uniqueness of the solution of (14) it is necessary to demonstrate that the homogeneous integral equation

$$0=\int_{-\infty}^{\infty}k(\tau)\psi(t-\tau)d\tau \qquad (15)$$

in a certain class of functions has only the trivial solution. Taking advantage of Titchmarsh's results [34], let us prove the following theorem.

THEOREM. Let there exist constants $c_1>c>0$ and $e^{c_1|t|}\psi(t) \in L_1(-\infty,\infty)$, $e^{-c|t|}k(t) \in L_2(-\infty,\infty)$. Then equation (15) has only the zeroth solution.

In the same manner as in theorem 146 of Titchmarsh's monograph

[34] it may be shown that the generalized Fourier integrals

$$K_+(\omega)=\frac{1}{\sqrt{2\pi}}\int_0^{\infty} k(t)e^{i\omega t}dt\ ,\quad K_-(\omega)=\frac{1}{\sqrt{2\pi}}\int_{-\infty}^{0} k(t)e^{i\omega t}dt\ ,$$

are functions regular in the band $b<\mathrm{Im}\,\omega<a$, where $c<a<c_1$ and $-c_1<b<-c$, with $K_+(\omega)=-K_-(\omega)$. Then the solution of equation (15) may be presented, for example, as

$$k(t)=\frac{1}{\sqrt{2\pi}}\int_{ib-\infty}^{ib+\infty} K_-(\omega)e^{-i\omega t}\,d\omega-\frac{1}{\sqrt{2\pi}}\int_{ia-\infty}^{ia+\infty} K_-(\omega)e^{-i\omega t}d\omega\ .$$

As proved, $K_-(\omega)$ is regular in the band $b<\mathrm{Im}\,\omega<a$ and under the conditions of the theorem it is regular for $\mathrm{Im}\,\omega<-c$. It implies that throughout the lower semiplane $\mathrm{Im}\,\omega<a$ $K_-(\omega)$ is a regular function and from Jordan's lemma it follows that $k(t)=0$.

Equation (13) may also be treated in a similar manner.

By way of illustration here is a case for which the foregoing is essential.

Exponential state will be taken to mean a cell population state in which $n(t)=C_0e^{pt}$.

The function $n(t)=C_0e^{pt}$ appears to possess generalized Fourier transformations: $N_+(\omega)$ for $\mathrm{Im}\,\omega>\mathrm{Re}\,p$ and $N_-(\omega)$ for $\mathrm{Im}\,\omega<\mathrm{Re}\,p$. Hence,

$$k_1(\tau)=C_0e^{p\tau}\left[p+\frac{\sqrt{2\pi}\,p^2\int_0^{\infty}e^{-px}F(x)dx}{1-\sqrt{2\pi}\,p\int_0^{\infty}e^{-px}F(x)dx}\right]=\frac{C_0e^{p\tau}}{\frac{1}{p}-\int_0^{\infty}e^{-px}F(x)dx}. \quad (16)$$

If $F(x)$ is the generation time distribution in an exponentially growing closed population with the generation coefficient (offspring mean) $\eta>1$, then (16) takes the form

$$k_1(t) = \frac{\eta-1}{p\eta} C_0 e^{pt} .$$

Hence and from formula (10) we obtain the familiar expression [23] for cell age distribution in a mitotic cycle

$$w(a,t) = \frac{p\eta}{\eta-1} e^{-pa} [1-F(a)] .$$

Thus, cell age distribution in an exponentially growing population is stationary.

The approach to solving equation (12) as outlined above presumes that the function $n(t)$ is known with absolute precision. However, if $n(t)$ is yielded by an experiment, i.e. instead of the accurate $n(t)$ values only an approximation $\hat{n}(t)$ involving some errors is available, constructing a solution becomes more complicated since the problem of solving an integral equation of the first kind will be ill-posed [33].

A regularizing algorithm [33] has been developed for approximate solution of such convolution-type equation on the basis of integral transformations.

Discussed in detail in monograph [19] is another approach to constructing a stable solution associated with a fairly reasonable assumption that the solution belongs to a class of finite functions bounded on certain symmetric interval.

It is rather seldom that need arises for solving integral equations (12) and (13) in analyzing cell kinetics. These equations may also be used merely as expressions enabling calculation of the functions $k_2(t)$ and $n(t)$ from the known rate of entry $k_1(t)$. In that way, for example, one may study the passage of a cohort of synchronized cells through mitotic cycle phases. A problem of that kind is discussed in the next section.

3.3. Investigation of Periodic Processes in Cell Kinetics

DNA synthesis and mitotic activity in the renewing tissues of an adult organism are known to undergo diurnal variations. Diurnal variations of a fraction of cells in some phase of the mitotic cycle may be caused by two factors:

1) periodical changes in phase length distribution with time;

2) partial synchronization of the population prior to the cells' entry into the phase.

Some authors are of the opinion that durations of the mitotic cycle phases vary with time of the day [5,38], whereas others [4, 6,15,26] believe that such variations can be neglected and that diurnal variations in the indices of labelled cells and mitoses are due solely to partial synchronization of the evolution of cells prior to their entry into the S-phase. The question of the possible trend in the mean length of the mitotic cycle phases will be taken up in Chapter IV. The approach taken in this section to analysis of periodic processes in cell kinetics is based on the assumption that such processes originate as a result of periodic changes in the rate of cells' passage to DNA synthesis and mitotic division. Thus, we shall assume that phase duration distribution is independent of the moment t' of the cell entry into the phase, i.e. $f(t-t',t')=f(t-t')$, and the signal that controls cell proliferation by blocking cells in the cycle is a periodic function with the basic period $T \simeq 24$ hr. Let the subphase sensitive to the effect of the controlling signal and the phase of mitosis be separated by some intermediary phase Z whose duration is distributed according to the density $f_Z(x)$.

Then the expected rate of cells' entry into the phase Z may be represented as the trigonometrical polynomial

$$k_{1Z}(t)=a_0 + \sum_{j=1}^{r} (a_j \cos \omega_j t + b_j \sin \omega_j t). \qquad (17)$$

The expected number of cells in the M-phase may be expressed in terms of the cells' entry into the Z-phase as

$$n_M(t)=\int_{-\infty}^{t} [1-F_M(t-t')dt' \int_{-\infty}^{t'} k_{1Z}(\tau) f_Z(t'-\tau)d\tau \ . \qquad (18)$$

Let us introduce the following designations:

$$\Psi(i\omega) = \int_0^\infty e^{-i\omega u} f(u)du \ , \tag{19}$$

$$\Phi(i\omega) = \int_0^\infty e^{-i\omega u} [1-F(u)]du \ , \tag{20}$$

The functions $\Psi(i\omega)$ and $\Phi(i\omega)$ defined in (19) and (20) will be called the first and the second frequency characteristics, respectively. Frequency characteristics were first used in cell kinetics analysis by Yakovlev [39,41]. A comprehensive discussion of applications of that approach is given in a later paper by Hartmann and Moller [17].

Let in (18)

$$k_{1Z}(\tau)=Ae^{i\omega\tau} \ , \text{ where } A=C_0e^{-i\zeta_0} \ ,$$

then, taking into account (19) and (20) the function $n_M(t)$ may be represented in the complex form as

$$n_M(t)=\Psi_Z(i\omega)\Phi_M(i\omega)Ae^{i\omega t} \ . \tag{21}$$

Now, if the function $\mathcal{F}_Z(i\omega)=\Psi_Z(i\omega)\Phi_M(i\omega)$ is represented in an exponential form, it will be easy to evaluate the change in the amplitude and the phase shift in variations occurring under the effect of the mechanisms which form the diurnal rhythm of cell proliferation processes. Separating in (21) the real part from the imaginary one, the function n_M may be written in the form similar to (17).

The following important property of integral operators of type (18) shows that there are serious grounds for the assumption of a stationary distribution of the Z- and M-phase durations. The integral operator of the type

$$B[t;x(\tau)] \equiv \int_{-\infty}^{t} x(\tau)g(t,\tau)d\tau = y(t) \tag{22}$$

transforms any periodic function $x(\tau)$ with an arbitrary period T into another periodic function $y(t)$ with the same period T when, and only when, its kernel depends solely on the difference of arguments, i.e. $g(t,\tau) \equiv g(t-\tau)$ [35].

Another possibility, however, should also be considered. Let the period T of the input signal $x(\tau)$ be fixed, and the operator $B[t;x(\tau)]$-periodic with the same period T, i.e. the weight function $g(t,\tau)$ would have the property of periodicity [29]:

$$g(t-T,\tau-T) = g(t,\tau) . \qquad (23)$$

Then in (22) we have

$$y(t) = \int_{-\infty}^{t-T} g(t-T,\tau)x(\tau)d\tau = y(t-T) .$$

Thus, the non-stationary system with weight function (23) does not change the period of input function provided the latter coincides with the weight function period. Note that for weight functions of the kind $g(t-\tau,\tau)$ condition (23) is equivalent to

$$g(t-\tau,\tau-T) = g(t-\tau,\tau) ,$$

i.e. the property of periodicity in that case relates only to the second argument. When polyharmonic signal (17) passes through such a system, the basic period T will remain unaltered.

If as an approximation of $f_Z(x)$ the Γ-distribution density is taken, then

$$\mathcal{F}_Z(i\omega) = \frac{\beta^{\alpha}}{i\omega(i\omega + \beta)^{\alpha}} \left[1 - \frac{\mu^{\lambda}}{(i\omega + \mu)^{\lambda}} \right] . \qquad (24)$$

Here

$$\frac{\alpha}{\beta} = \bar{\tau}_Z , \quad \frac{\sqrt{\alpha}}{\beta} = \sigma_Z ; \quad \frac{\lambda}{\mu} = \bar{\tau}_M , \quad \frac{\sqrt{\lambda}}{\mu} = \sigma_M ;$$

$\bar{\tau}$ is the mean phase duration and σ is the corresponding standard deviation.

The amplitude frequency characteristic of the Z-phase (of system (18)) is obtained from (24) after separating the real part from the imaginary one:

$$|\mathcal{F}_Z(i\omega)| = \frac{\sqrt{1+\left(1+\frac{\omega^2}{\mu^2}\right)^{\lambda} - 2\left(1+\frac{\omega^2}{\mu^2}\right)^{\lambda/2}\cos\left[\lambda \operatorname{arctg}\left(-\frac{\omega}{\mu}\right)\right]}}{\omega\sqrt{\left(1+\frac{\omega^2}{\beta^2}\right)^{\alpha}\left(1+\frac{\omega^2}{\mu^2}\right)^{\lambda}}} . \quad (25)$$

The phase frequency characteristic of the Z-phase will be expressed by the formula

$$\operatorname{tg}(\arg\mathcal{F}_Z(i\omega)) = \frac{\mu^{\lambda}\rho^{\lambda}\cos(\lambda\varphi_0+\lambda\psi_0)-(\mu^2+\omega^2)^{\lambda}\cos(\alpha\varphi_0)}{\mu^{\lambda}\rho^{\lambda}\sin(\alpha\varphi_0+\lambda\psi_0) - (\mu^2+\omega^2)^{\lambda}\sin(\alpha\varphi_0)} , \quad (26)$$

where

$$\rho = \sqrt{\mu^2+\omega^2}\,, \quad \varphi_0 = \operatorname{arctg}\left(-\frac{\omega}{\beta}\right), \quad \psi_0 = \operatorname{arctg}\left(-\frac{\omega}{\mu}\right).$$

Let us now write out separately the analytical experessions for the modulus and the argument of the first and second frequency characteristics of any cell cycle phase. Let γ and m be the scale and form parameters of Γ-distribution, respectively, the latter approximating the particular phase distribution length, then

$$|\Psi(i\omega)| = \left(1 + \frac{\omega^2}{\gamma^2}\right)^{-m/2} \tag{27}$$

$$\arg \Psi(i\omega) = m \operatorname{arctg} \left(-\frac{\omega}{\gamma}\right) ; \tag{28}$$

$$|\Phi(iw)| = \frac{\sqrt{1 + \left(1 + \frac{\omega^2}{\gamma^2}\right)^m - 2\left(1 + \frac{\omega^2}{\gamma^2}\right)^{m/2} \cos(m\varphi_0)}}{\omega \sqrt{\left(1 + \frac{\omega^2}{\gamma^2}\right)^m}} , \tag{29}$$

$$\arg \Phi(i\omega) = \operatorname{arctg} \left\{ - \frac{\gamma^m \rho^m \cos(m\varphi_0) - (\gamma^2+\omega^2)^m}{\gamma^m \rho^m \sin(m\varphi_0)} \right\} ; \tag{30}$$

where

$$\rho = \sqrt{\gamma^2+\omega^2} , \quad \varphi_0 = \operatorname{arctg} \left(-\frac{\omega}{\gamma}\right) .$$

Asymptotics of the function $|\mathcal{F}_Z(i\omega)|$ can be obtained directly from expression (25): with $\omega \to 0$, $|\mathcal{F}_Z(i\omega)| \to \bar{\tau}_M$; with $\omega \to \infty$, $|\mathcal{F}_Z(i\omega)| \to 0$. This behaviour of the function $|\mathcal{F}_Z(i\omega)|$ is presumably due to the asymptotic properties of the first and second amplitude frequency characteristics $|\Psi_Z(i\omega)|$ and $|\Phi_M(i\omega)|$, which appear in the expression for $|\mathcal{F}_Z(i\omega)|$ as cofactors. Indeed,

$$\lim_{\omega\to\infty} |\Psi(i\omega)| = 1; \quad \lim_{\omega\to\infty} |\Phi(i\omega)| = \bar{\tau} ; \tag{31}$$

and on by virtue of the Riemann-Lebesque theorem it may be

said that

$$\lim_{\omega\to\infty} |\Psi(i\omega)| = 0 \quad \text{and} \quad \lim_{\omega\to\infty} |\Phi(i\omega)| = 0 .$$

The results of computer-aided numerical studies of the dependence of the $|\Psi(i\omega)|$ and $|\Phi(i\omega)|$ functions on the frequency ω presented in Table 1 show the trend of the functions towards their limiting values in accordance with (31). The form of relationship between frequency characteristics and frequency ω is also determined by values of the temporal parameters $\bar{\tau}$ and σ of the cell cycle phase under consideration. For a more comprehensive treatment of the problem we shall perform a reduction to a population of ideal cells whose residence time in a given cycle phase is constant and equal to $\bar{\tau}$. In that case we have

$$\Psi_{id}(i\omega) = e^{-i\omega\bar{\tau}} \; ; \; \Phi_{id}(i\omega)=\frac{1 - e^{i\omega\bar{\tau}}}{i\omega};$$

$$|\Psi_{id}(i\omega)| = 1 \; ; \qquad |\Phi_{id}(i\omega)|=\frac{\bar{\tau}\left|\sin\left(\frac{\omega\bar{\tau}}{2}\right)\right|}{\frac{\omega\bar{\tau}}{2}};$$

$$\arg \Psi_{id}(i\omega)=-\omega\bar{\tau} \; , \; \arg \Phi_{id}(i\omega)=-\frac{\omega\bar{\tau}}{2} .$$

Thus, as one would expect, $\Phi_{id}(i\omega)$ is a typical frequency characteristic of the current integration operator, i.e.

$$n(t) = \int_{t-\bar{\tau}}^{t} k_1(x)dx . \qquad (32)$$

Transform (32) possesses selectivity to low frequencies and is used extensively in problems of stochastic process smoothing. The modulus of the second frequency characteristic of an ideal population with $\bar{\tau}$ = 10 hr is shown in Fig.1. Indeed, with low values of the variation coefficient $V=\frac{\sigma}{\bar{\tau}}$ for cycle phase

Table 1.

Relationship between the functions $|\Phi(i\omega)|$ and $|\Psi(i\omega)|$ and the frequency ω with fixed values of the temporal parameters $\bar{\tau}=9$ hr; $\sigma=1.8$ hr

j	$\|\Phi(i\omega)\|$ with $\omega=\pi 2^{1-j}$	$\|\Psi(i\omega)\|$ with $\omega=\pi 2^{1-j}$
-10	0.0001	0.0000
-9	0.0003	0.0000
-8	0.0006	0.0000
-7	0.0012	0.0000
-6	0.0025	0.0000
-5	0.0050	0.0000
-4	0.0099	0.0000
-3	0.0199	0.0000
-2	0.0398	0.0000
-1	0.0796	0.0000
0	0.1591	0.0000
1	0.3183	0.0000
2	0.6177	0.0312
3	0.9150	0.3824
4	4.4589	0.7809
5	7.6293	0.9396
6	8.6405	0.9845
7	8.9090	0.9961
8	8.9772	0.9990
9	8.9943	0.9998
10	8.9986	0.9999

duration the amplitude frequency characteristic of a real population defined by expression (29) may differ but little from $|\Phi_{id}(i\omega)|$ and, consequently, it can possess identical resonance frequencies. In fact it is evident from Fig.1 that if $\sigma=0.4$ hr, then with $\bar{\tau}=10$ hr the amplitude frequency characteristics $|\Phi(i\omega)|$

and $|\Phi_{id}(i\omega)|$ will practically coincide. However, with σ=2.5 hr

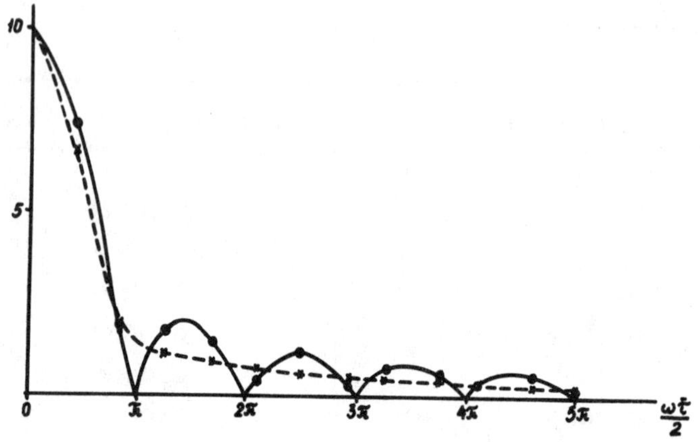

Figure 1. Relation between the amplitude frequency characteristics $|\Phi(i\omega)|$ and $|\Phi_{id}(i\omega)|$ with the mean cycle phase duration of 10 hr.

On the abscissa - the variable $\frac{\omega\bar{\tau}}{2}$ values; on the ordinate - amplitude frequency characteristic values (hr).

The solid line represents $|\Phi_{id}(i\omega)|$, the dots indicate individual $|\Phi(i\omega)|$ values calculated with σ=0.4 hr and the dashed line shows the behaviour of $|\Phi(i\omega)|$ with σ=2.5 hr.

the type of the $|\Phi(i\omega)|$ characteristic changes, acquiring features more specific to populations of real cells. This may affect the results of frequency analysis of cell kinetics curves.

By combining amplitude and phase frequency characteristics of mitotic cycle periods (27),(28),(29) and (30) one can trace the passage through the cycle of the "synchronization wave" generated by mechanisms of the diurnal rhythm of cell proliferation processes. Let us consider experimentally observed diurnal variations in the number of cells (with a constant total number of cells in the system use may be made of cycle phase indices) in DNA synthesis and mitosis. As previously, we are of the opinion that such variations are due to periodic changes in the rate of cells' entry into the S-phase. In that case, comparing the amplitude range of variations of labelled cells' index with the range of mitotic index variations one may obtain evidence in favour of the presence or lack of feedback influence in the G_2-period of the mitotic cycle. It is also possible to study the balance of cells in the neighbouring cycle phases by comparing the rates of cells' entry into the phases. Using the first frequency characteristic of the phase $\Psi(i\omega)$, the required rate transformations may be performed by the formula

$$k_2(t) = a_0 + \sum_{j=1}^{r} |\Psi(i\omega_j)| \Big\{ a_j \cos[\omega_j t + \arg\Psi(i\omega_j)] +$$

$$b_j \sin[\omega_j t + \arg\Psi(i\omega_j)] \Big\} .$$

Thus the case when the rate of cells' entry into a given phase of the mitotic cycle has the form of (17) is simple to treat, and the frequency properties of cell cycle phases are of interest all their own. Thus, for instance, if the dynamic structure of all the cycle phases under review is stationary, it is possible to estimate the variance of the phase $S+G_2$. Let us consider two identical in frequency harmonics from curves for diurnal variations in the number of cells in the S- and M-phases. The amplitudes of the harmonics (A_j for the S-phase and B_j for the M-phase) are connected by the apparent relation

$$A_j |\mathcal{F}_{S+G_2}(i\omega_j)| = B_j |\Phi_S(i\omega_j)| \tag{33}$$

Substituting in (33) formulas (25) and (29) we obtain

$$\xi_j = \frac{1}{\Theta_j} \sqrt{\left(1 + \frac{\omega_j^2}{\beta^2}\right)^{\alpha}} , \tag{34}$$

where $\xi_j = A_j/B_j$; $\Theta_j = |\Phi_M(i\omega_j)|/|\Phi_S(i\omega_j)|$; α and β are the

$S+G_2$-phase parameters. Rewrite (34) as

$$\xi_j^2 \, \Theta_j^2 = (1 + \frac{\omega_j^2}{\beta^2})^{\alpha} \, . \tag{35}$$

Selecting the frequency ω_j so that $\frac{\omega_j^2}{\beta^2} \ll 1$ and expanding (35) into a binomial series we arrive at the final approximate formula for the $S+G_2$-phase variance

$$\sigma^2_{S+G_2} \simeq \frac{\xi_j^2 \, \Theta_j^2 - 1}{\omega_j^2} \, ,$$

where the variable $\xi_j \Theta_j$ in its sense is always greater than one.

Note that the ratio of the constant component of the curve for diurnal variation in the number of labelled cells to the corresponding constant component of diurnal variation in the number of mitoses (let it be denoted by χ) is of the form

$$\chi = \bar{\tau}_S \, / \, \bar{\tau}_M \, . \tag{36}$$

Hence, knowing $\bar{\tau}_M$ one can by means of (36) evaluate $\bar{\tau}_S$ and vice versa.

In conclusion it may be said that the substantial variability of experimental evidence concerning the diurnal rhythm of indices of labelled cells and mitoses creates considerable complications in performing analysis, accounting for very broad confidence intervals constructed for amplitudes by means of familiar statistical methods [18,30].

3.4. Basic Integral Equations for Unsteady State Cell Kinetics

The starting point in discussing methods for mathematical description of transient processes in cell kinetics will be a closed population of cells. Let us find such a generalized form of equation for the mean size of a population in the case of $\eta > 1$, i.e. a renewal equation of the form

$$\mathbb{E}\{N(t)\} = \eta\int_0^t \mathbb{E}\{N(t-\tau)\}dF(\tau)+1-F(t), \qquad (37)$$

which would provide a means for modelling the dynamics of a closed population with an arbitrary initial age distribution $w(a,0)$. The designations used are those introduced in Section 3.1. Let us rewrite equation (12) as

$$\mathbb{E}\{N(t)\} = \int_0^t \mathbb{E}\{K_1(\tau)\}[1-F(t-\tau)]d\tau+v(t) , \qquad (38)$$

where the function

$$v(t) = \int_{-\infty}^0 \mathbb{E}\{K_1(\tau)\}[1-F(t-\tau)]d\tau \qquad (38')$$

will be referred to in the sequel as a transient process. From (38) we obtain

$$\mathbb{E}\{N(t)\} = v(t) + \int_0^t \mathbb{E}\{K_1(\tau)\}d\tau - \int_0^t \left(\int_0^{t-\tau} \mathbb{E}\{K_1(x)\}dx \right) dF(\tau) . \qquad (39)$$

Seeing that under consideration is a closed population with the generation coefficient η for reasons of balance we have

$$\mathbb{E}\{N(t) - N(0)\} = \frac{1}{\eta} \int_0^t \mathbb{E}\{K_1(x)\}dx . \qquad (40)$$

Substituting (40) in (39), we obtain an equation generalizing equation (38) for the case when at the initial moment $t=0$ there are $N(0)$ cells in a population. That equation is of the form

$$\mathbb{E}\{N(t)\} = \eta\mathbb{E}\{N(0)\}[1-F(t)] - v(t) + \eta \int_0^t \mathbb{E}\{N(t-\tau)\}dF(\tau) . \qquad (41)$$

Now it remains to ascertain in more detail the meaning of the expression for $v(t)$. Set $t=0$ in formula (10). Thus, the initial cell age distribution density is determined by the expression

$$w(a,0) = \mathbb{E}\{K(-a)\}[1-F_1(a)]/\mathbb{E}\{N(0)\} . \tag{42}$$

Substituting (42) in (38'), we obtain

$$v(t) = \mathbb{E}\{N(0)\}\int_0^{\infty} w(\tau,0)\frac{[1-F(t+\tau)]}{[1-F(\tau)]}\, d\tau . \tag{43}$$

In view of (43) equation (41) takes the form

$$\mathbb{E}\{N(0)\} = \eta\mathbb{E}\{N(0)\}[1-F(t)] + \eta\int_0^t \mathbb{E}\{N(t-\tau)\}dF(\tau) - \mathbb{E}\{N(0)\}\int_0^{\infty} \frac{[1-F(t+\tau)]}{[1-F(\tau)]}\, w(\tau,0)d\tau \tag{44}$$

The function $[1-F(t+\tau)]/[1-F(\tau)]$ in (43) is the conditional probability that a cell reaches an age greater than $t+\tau$, provided the age τ has already been reached. For this reason equation (44) may be derived on the strength of purely probabilistic considerations. The procedure of deducing equation (44) given above has been chosen because of its simplicity. Martinez [24] obtained in a somewhat different way a similar generalization of the Bellman-Harris equation, i.e. equation (37) with $\eta=2$:

$$\mathbb{E}\{N(t)\}=2\mathbb{E}\{N(0)\}[1-F(t)] + 2\int_0^t \mathbb{E}\{N(t-\tau)\}dF(\tau) - \int_{-\infty}^{0} \frac{[1-F(t-\tau)]}{[1-F(-\tau)]}\, dH(\tau) ,$$

where $H(-\tau)$ is the mean number of cells of an age not above τ and is the function of bounded total variation.

For transitive cell populations equations may also be derived which make it possible to discern the effect of the initial state of a population on the kinetics observed after $t=0$. Rewrite equations (12) and (13) as follows

$$\mathbb{E}\{N(t)\} = \int_0^t \mathbb{E}\{K_1(\tau)\}[1-F(t-\tau)]d\tau + \int_0^\infty \mathbb{E}\{K_1(-\tau)\}[1-F(t+\tau)]d\tau$$

$$\mathbb{E}\{K_2(t)\} = \int_0^t \mathbb{E}\{K_1(\tau)\}f(t-\tau)d\tau + \int_0^\infty \mathbb{E}\{K_1(-\tau)\}f(t+\tau)d\tau \ .$$

Taking advantage of expression (10) let us put these equations into a more convenient form

$$\mathbb{E}\{N(t)\} = \int_0^t \mathbb{E}\{K_1(\tau)\}[1-F(t-\tau)]d\tau + \mathbb{E}\{N(0)\}\int_0^\infty \frac{[1-F(t+\tau)]}{[1-F(\tau)]} w(\tau,0)d\tau \quad , \tag{45}$$

$$\mathbb{E}\{K_2(t)\} = \int_0^t \mathbb{E}\{K_1(\tau)\}f(t-\tau)d\tau + \mathbb{E}\{N(0)\}\int_0^\infty \frac{f(t+\tau)}{[1-F(\tau)]} w(\tau,0)d\tau \ . \tag{46}$$

With certain suitable restrictions on $w(\tau,0)$ and $f(\tau)$ the second terms of the equations (i.e. the transient processes $v(t)$), non-increasing with t, tend to zero at $t\to\infty$, which implies asymptotic stability. Integrating by parts, (45) is reduced to the integral equation of the second kind

$$\mathbb{E}\{N(t)\} = r(t) - \int_0^t r(t-\tau)dF(\tau) + v(t), \tag{47}$$

where $r(t) = \int_0^t \mathbb{E}\{K_1(\tau)\}d\tau$ and $v(t)$ is defined by expression (43). Considering that $F(x)$ is an absolutely continuous function, rewrite (47) in the form

$$\mathbb{E}\{N(t)\} = r(t) - \int_0^t r(t-\tau)f(\tau)d\tau + v(t) \ . \tag{48}$$

In the case of an arbitrary initial moment $t_0 \neq 0$ equation (45) takes the form

$$\mathbb{E}\{N(t)\}=\int_{t_0}^{t}\mathbb{E}\{K_1(\tau)\}[1-F(t-\tau)]d\tau + v(t;t_0) , \qquad (49)$$

where

$$v(t;t_0) = \int_{-\infty}^{t_0} \mathbb{E}\{K_1(\tau)\}[1-F(t-\tau)\}d\tau =$$

$$\mathbb{E}\{N(t_0)\} \int_0^{\infty} \frac{[1-F(t-t_0+\tau)]}{[1-F(\tau)]} w(\tau,t_0)d\tau . \qquad (49')$$

Equation (48) can also be put into a similar form

$$\mathbb{E}\{N(t)\} = r(t;t_0) - \int_{t_0}^{t} r(\tau;t_0)f(t-\tau)d\tau + v(t;t_0) , \qquad (50)$$

where the function $r(t;t_0)=\int_{t_0}^{t}\mathbb{E}\{K_1(\tau)\}d\tau, t>t_0$, is the integral influx (flow, stream) of cells into a given population determined for an arbitrary t_0.

In considering further the basic cell kinetics equation it is expedient to take into account to a larger extent the existing experimental indices. For this reason instead of the absolute number of cells $N(t)$ we shall turn to the phase index, defining it as the ratio between the expected number of cells in a given cell cycle phase and the expected total number of cells in the population $\mathbb{E}N_\Sigma(t)$ at the moment t. Thus, for the ith cycle phase its index is defined as

$$I_i(t) = \mathbb{E}\{N_i(t)\}/\mathbb{E}\{N_\Sigma(t)\}.$$

This definition of the index $I_i(t)$ requires substantiation. In biological experiment the frequencies $N_i(t)/N_\Sigma(t)$ are observed which are used as estimators of the probability

$$p_i = \mathbb{P}\left\{ \begin{array}{l} \text{randomly selected cell pertaining} \\ \text{to the i-th cycle phase at the moment t} \end{array} \right\} ;$$

Considering the random selection of cells pertaining to a given cycle phase from the standpoint of the Bernoulli trials, we come to the following representation of the function $N_i(t)$

$$N_i(t) = \xi_1 + \xi_2 + \dots + \xi_{N_\Sigma(t)} ,$$

where $\{\xi_k\}_{k=1}^{\infty}$ is the sequence of independent identically distributed random variables with distribution

$$P\{\xi_k=1\}=p_i(t) , \quad P\{\xi_k=0\}=1-p_i(t).$$

Assuming $N_\Sigma(t)$ to be a random variable "independent of the future" and using the Wald's equality, we have

$$\mathbb{E}\{N_i(t)\}=\mathbb{E}\{\xi_k\}\mathbb{E}\{N_\Sigma(t)\}=p_i(t)\mathbb{E}\{N_\Sigma(t)\} .$$

Thus

$$p_i(t) = \frac{\mathbb{E}\{N_i(t)\}}{\mathbb{E}\{N_\Sigma(t)\}} .$$

It is also convenient to introduce the notion of modified phase index $\tilde{I}_i(t)=\mathbb{E}\{N_i(t)\}/\mathbb{E}\{N_\Sigma(0)\}$, which characterizes, same as $I_i(t)$, the fraction of cells in the i-th phase at the moment t, but now in relation to the <u>initial</u> total number of cells in the system. Finally, let us introduce one more kinetic index $q_i(t)$ [42]:

$$q_i(t) = \int_0^t \mathbb{E}(K_{1i}(\tau)\}d\tau/\mathbb{E}\{N_\Sigma(0)\} = r_i(t)/\mathbb{E}\{N_\Sigma(0)\} , \qquad (51)$$

which will be referred to as the q-index. It is apparent that the q-index characterizes the integral flow of cells into the i-th phase related to the initial mean size of the entire cell system. Similarly it is possible to determine the index

$$q_i(t,t_0) = r_i(t,t_0)/\mathbb{E}\{N_\Sigma(t_0)\} , \qquad (52)$$

which will be used in the next chapter in the analysis of labelled mitoses curve.

Dividing both sides of equation (48) by the initial expected number of cells $\mathbb{E}\{N_\Sigma(0)\}$ we come to the integral equation of the second kind for the function $q(t)$

$$\tilde{I}(t) = q(t) - \int_0^t q(\tau)f(t-\tau)d\tau + \tilde{I}(0)\int_0^\infty \frac{[1-F(t+\tau)]}{[1-F(\tau)]} w(\tau,0)d\tau. \quad (53)$$

Taking into account that $\tilde{I}(0)=I(0)$ and introducing the designation

$$\tilde{v}(t) = I(0)\int_0^\infty \frac{[1-F(t+\tau)]}{[1-F(\tau)]} w(\tau,0)d\tau , \quad (53')$$

rewrite (53) in the following compact form

$$\tilde{I}(t) - \tilde{v}(t) = q(t) - \int_0^t q(\tau)f(t-\tau)dt . \quad (54)$$

If $q(t)$ is represented as the difference of two functions, i.e. $q(t)=q_1(t)-q_2(t)$, (54) will be equivalent to the system of two equations

$$\tilde{I}(t) = q_1(t) - \int_0^t q_1(\tau)f(t-\tau)dt \quad (55)$$

$$\tilde{v}(t) = q_2(t) - \int_0^t q_2(\tau)f(t-\tau)d\tau , \quad (56)$$

the former representing the effect on the cell kinetics of the free term - $\tilde{I}(t)$ and the latter - the effect of initial conditions (transient process $\tilde{v}(t)$). Equations (54),(55) and (56) are particular variants of the familiar integral equation of the renewal theory. Relying on some results obtained for that equation [3,11,12,16] and taking into consideration the real properties of the functions $\tilde{I}(t)$, $f(t)$ and $\tilde{v}(t)$ we can obtain data on solutions of equations (55) and (56) beneficial to applications without explicitly defining the analytical form of these functions.

It is natural to believe that for any interval $[0,t_0]$ the following conditions are fulfilled:

$$0 \le \tilde{I}(t) \le c_1 = \tilde{I}_{max} \; ; \qquad \int_0^{t_0} \tilde{I}(\tau)d\tau < \infty \; ;$$

$$0 \le f(t) \le c_2 = f(x_{mod}) \; ; \quad \int_0^{t_0} f(\tau)d\tau \le 1 \; ; \qquad (57)$$

x_{mod}=mode of distribution $F(x)$. The function $\tilde{v}(t)$ described in (53) has the following properties:

(1) As estimate $1-F(t+x) \le 1-F(x)$ and the condition of normalization

$$\int_0^{\infty} w(a,o)da=1$$

indicate, the integral which determines $\tilde{v}(t)$ converges uniformly at all t, and $\tilde{v}(t) \le I(0)$. At $t=0$ $\tilde{v}(t)=I(0)$, while at $t>0$, $\tilde{v}(t) \le I(0)$.

(2) The function $\tilde{v}(t)$ is continuous with respect to t, non-increasing monotonically as t increases and tending to zero as $t \to \infty$.

Thus, the conditions (57) may be complemented by, at least, two more conditions :

$$0 \le \tilde{v}(t) \le I(0) \; ; \int_0^{t_0} \tilde{v}(\tau)d\tau < \infty \, . \qquad (58)$$

The whole complex of conditions (57) and (58) provides for a solution existence for each of the equations (54), (55) and (56) which is the only one, bounded, non-negative and integrable in every finite interval $[0,t_0]$. Moreover, if the left-hand sides of the equations are continuous within $[0,t_0]$, their solutions are continuous too. From the theorem of uniqueness of non-negative solution for the case $\tilde{I}(t) \ge \tilde{v}(t)$ it follows that $q_1(t) \ge q_2(t)$. Besides, it is possible to claim that in the case of induced cell proliferation $q_1(t)-q_2(t) \ge z(t)$, where $z(t)$ is a monotonically non-decreasing function [43].

The effect of initial conditions in the case of large t values is estimated by asymptotic behaviour of the solution of equation

(56). The function $\tilde{v}(t)$, non-increasing and integrable on the interval $[0,\infty)$, is "directly Riemann integrable" on that interval [12].

Hence, from the principal renewal theorem follows the asymptotic equality

$$q_2(t) \sim \frac{1}{\bar{\tau}} \int_0^t \tilde{v}(x)dx \ , \ t\to\infty \ , \tag{59}$$

where

$$\bar{\tau} = \int_0^\infty [1-F(x)]dx \ .$$

The properties of the mathematical model of unsteady-state cell kinetics just discussed may to some extent be used in analyzing such systems with induced or stimulated proliferation whose initial (pre-stimulation) level of proliferative activity is too high to warrant neglecting a priori the effect of initial conditions and using at once equation (55) for calculating $q(t)$. In certain special cases the function $\tilde{v}(t)$ may be expressed in the explicit form and substituted into the solution of the basic integral equation of semistochastic cell kinetics (54). In such cases a separate study of the significance of the $\tilde{v}(t)$ contribution to cell kinetics is no longer essential and the estimates given in this section lose their importance for solving practical problems. The construction of the transient processes $\tilde{v}(t)$ under certain specific conditions of cell proliferation (e.g., when the system prior to receiving a proliferative stimulus is in the strict sense steady state or in the regime of diurnal rhythm of cell proliferation) will be dealt with in Section 3.6.

Solution of the renewal equation has already been given in the preceding chapter. If $f(x)$ is the density of Γ-distribution with the form parameter α and the scale parameter β, and $\alpha \geq 1$, the solution of equation (54) is of the form

$$q(t)=\tilde{I}(t)-\tilde{v}(t)+\sum_{k=1}^{\infty} \frac{\beta^{\alpha k}}{\Gamma(\alpha k)} \int_0^t [\tilde{I}(t-\tau)-\tilde{v}(t-\tau)]e^{-\beta\tau}\tau^{\alpha k-1}d\tau \ . \tag{60}$$

Thus, with the defined transient process $\tilde{v}(t)$ we are in a position to construct the kinetic index $q(t)$.

3.5. Construction of the q-Index of the S-Phase in a Special Case

The present section is concerned with certain aspects of practical q-index applications to analysis of systems with induced or stimulated cell proliferation (SISP) characterized by an initial level of proliferative processes which is so low that we may at once assume $\tilde{v}(t) \equiv 0$ for any mitotic cycle phase. This kind of SISP is typified by the regenerating liver of the mammals. Let us define the q-index for the S-phase of the mitotic cycle

$$q_S(t) = \frac{r_S(t)}{\mathbb{E}\{N_\Sigma(0)\}} , \tag{61}$$

and in view of $v_S(t) \equiv 0$ expression (60) takes the form

$$q_S(t) = \tilde{I}_S(t) + \sum_{k=1}^{\infty} \frac{\beta^{\alpha k}}{\Gamma(\alpha k)} \int_0^t \tilde{I}_S(t-\tau) e^{-\beta\tau} \tau^{\alpha k-1} d\tau , \tag{62}$$

where $\tilde{I}_S(t)$ is the modified index of labelled cells following pulse administration of ^{3}H-thymidine. In the discussion that follows we shall consider only expectations of random processes, using again for their designation the lower case letters $n(t)$, $k_1(t)$ and $k_2(t)$, and omitting the symbol $\mathbb{E}\{\cdot\}$.

Practical realization of formula (62) presupposes the existence of procedures for determining the duration parameters of the S-phase: α_S and β_S (or $\bar{\tau}_S$ and σ_S) and the index $\tilde{I}_S(t)$. Various experimental evidence indicates that in SISP the labelled cells index $I_S(t)$ usually increases much before the mitotic index $I_M(t)$ starts to rise. In that case, up to the commencement of a marked rise in mitotic activity, the $q_S(t)$ plot should coincide with the experimental curve for the labelled cells index with continuous administration of ^{3}H-thymidine – $I_S^c(t)$, while the modified index $\tilde{I}_S(t)$ may be substituted in calculations by the index $I_S(t)$. Thus, at the initial stage of development of the proliferative reaction: $t \in [0,t_0]$, characterized by the condition $n_\Sigma(t) \simeq n_\Sigma(0)$, the parameters α and β may be sought, for example, from the minimization of the function

$$\Theta(\bar{\tau},\sigma) = \sum_i \theta_i [q_S(t_i;\bar{\tau},\sigma) - I_S^c(t_i)]^2 . \qquad (63)$$

After substituting in (62) the experimentally determined function $\tilde{I}_S(t)=I_S(t)$ direct optimization of the parameters $\bar{\tau}_S$ and σ_S is possible. By way of example below is estimation of temporal parameters of the S-phase in fibroblast culture from the human lung stimulated to proliferation by addition of serum. The experimental data have been borrowed from Ellem and Mironescu [9]. Estimation of the parameters was performed by fitting the function $q_S(t)$ to the initial portion of the experimental curve $I_S^c(t)$ (the portion corresponding to a period not exceeding 22 hr after stimulation). Fig.2 demonstrates the relationship between the function $\Theta(\bar{\tau},\sigma)$ and the temporal parameters of the S-phase in the range of real values with $\theta_i=1$. Determination of the global minimum of $\Theta(\bar{\tau},\sigma)$ according to the data in Fig.2 yields the following estimates: $\bar{\tau}_S=10$ hr, $\sigma_S=1,5$ hr. Thus obtained, the S-phase parameters estimates may now be substituted in expression (62), and the $q_S(t)$ index may be calculated on the entire interval of observation, provided the form of the function $\tilde{I}_S(t)$ is known. The index $\tilde{I}_S(t)$ can be constructed from experimental data on $I_S(t)$. For this it is necessary to introduce a correction for cell multiplication, taking into account the mitotic index $I_M(t)$.

The concept of modified index was first introduced by Fabrikant [10] in studying the kinetics of cell proliferation in the regenerating rat liver. The method for constructing the $\tilde{I}_S(t)$ curve given by the author is based on the assumption that the duration of mitosis is identical for all cells and equal to 1 hr. We make use of another technique based, in contrast to Fabrikant's method, on the well-known facts relating to very great variability of the duration of mitosis in individual cells [28,40].

With very great variability and a low mean value of the

duration of mitosis (compared to the mean length of the whole cycle) it is reasonable to set

$$\frac{dn_{\Sigma}}{dt} = n_M(t)/\bar{\tau}_M , \qquad (64)$$

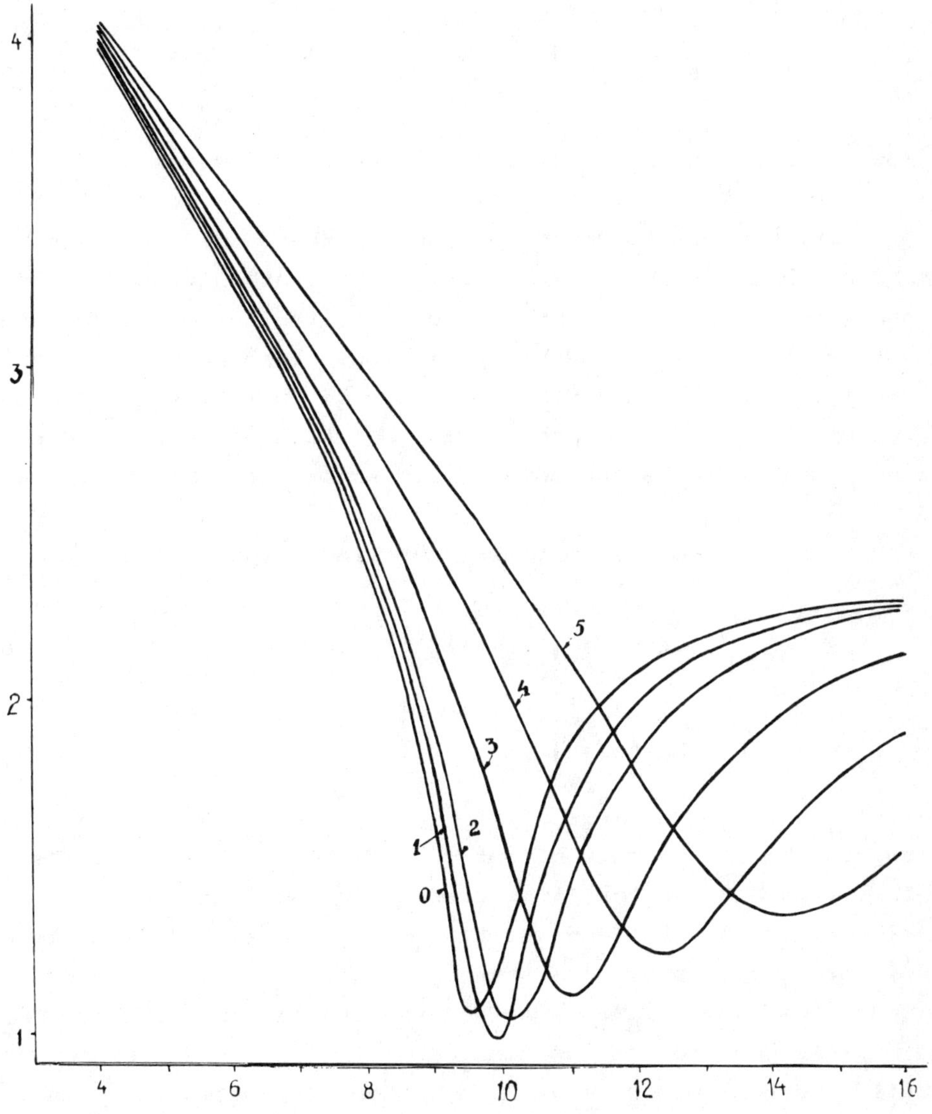

Figure 2. Optimization procedure for the temporal parameters $\bar{\tau}_S$ and σ_S in a cell culture stimulated to proliferation (according to data from [9]). On the abscissa - the mean duration of the S-phase; On the ordinate - lg $\Theta(\bar{\tau},\sigma)$. The relationship between lg $\Theta(\bar{\tau},\sigma)$ and $\bar{\tau}$ is shown at a fixed value of the variation coefficient $V=\sigma/\bar{\tau}$.

i.e. to neglect the age dependence of the process of cell exit from a phase of mitosis in the general variation of the size of the whole cellular system. Indeed, expression (64) implies that every cell entering into mitosis must complete it, giving rise to two daughter cells. Following directly from (64) is the way for determining $\tilde{I}_S(t)$:

$$\tilde{I}_S(t)=I_S(t)\exp\left\{\frac{1}{\bar{\tau}_M}\int_0^t I_M(x)dx\right\}. \tag{65}$$

Now, by substituting (65) in (62) the construction of the index $q_S(t)$ may be finally completed.

A fraction of cells entering the S-phase by the moment t comprises both such cells that begin synthesizing DNA for the first time after the application of the stimulus and those resynthesizing it. The index $q_S(t)$ thus represents together the processes of initial and repeated entry of cells into the S-phase of the mitotic cycle. A special consideration of these processes extends substantially the potentialities of kinetic analysis of SISP.

Let us introduce the following auxiliary kinetic indices

$$L_S(t) = \int_0^t l_S(t')dt'/n_\Sigma(0) , \tag{66}$$

$$H_S(t) = \int_0^t u_S(t')dt'/n_\Sigma(0) , \tag{67}$$

where $l_S(t)$ is the rate of entry into the S-phase of cells synthesizing DNA for the first time after application of a proliferative stimulus and $u_S(t)$ is the rate of repeated cell entry into the S-phase.

Thus, the index $q_S(t)$ may be represented by the sum $L_S(t)$ and $H_S(t)$. Compare the function $L_S(t)$ to the experimental labelled cell index curve with continuous (repeated or constant) labelling which for SISP is given by the following expression

$$I_S^c = \frac{\int_0^t l_S(t')dt' + n_\Sigma(t) - n_\Sigma(0)}{n_\Sigma(t)} \quad . \qquad (68)$$

Formula (68) is based on the reasoning that with continuous labelling all the cells that have divided within the observation period are labelled, and the inclusion by a cell of a labelled precursor for the second (or subsequent) time does not contribute to the value of the $I_S^c(t)$ index. However, this should be regarded only as a likely supposition.

From (66) and (68) follows the formula connecting the indices $L_S(t)$ and $I_S^c(t)$:

$$L_S(t) = [I_S^c(t) - 1] \frac{n_\Sigma(t)}{n_\Sigma(0)} + 1 \quad . \qquad (69)$$

But since usually $n_\Sigma(t) \geq n_\Sigma(0)$, so

$$L_S(t) \leq I_S^c(t) \; . \qquad (70)$$

The equality is valid in the absence of cell divisions and cell death, i.e. with $n_\Sigma(t) = n_\Sigma(0)$, or with $I_S^c(t) = 1$.

In experimental studies use is often made of the notion of "proliferative pool" which, as applied to SISP, is understood to mean the fraction of the initial number of cells $n_\Sigma(0)$ involved in proliferation reaction. Proliferative pool is frequently calculated from the plateau height in the $I_S^c(t)$ curve. It should be readily apparent that, in defining the notion of proliferative pool, the fixed time T of the cells' contact with a labelled precursor should always be taken into account.

In SISP it is reasonable to set the time T of constant cell contact with label equal to the whole period of existence of induced proliferation processes (a sufficiently long period of observation of unsteady-state cell kinetics). Then proliferative pool will be defined by the formula

$$L_S(T) = \frac{\int_0^T 1_S(t)dt}{n_\Sigma(0)}.$$

From inequality (70) it follows that in the general case the index $L_S(T)$ cannot be calculated from the $I_S^c(t)$ curve alone. Reverting to formula (69) and equation (64) the following procedure may be proposed for assessing the variable $L_S(T)$ in SISP:

$$L_S(T) = 1 + [I_S^c(T) - 1]\exp\left\{\frac{1}{\bar{\tau}_M}\int_0^T I_M(t)dt\right\}. \qquad (71)$$

The use of the $L_S(T)$ index yields important information on the regulation of cell proliferation reactions when it is essential to give general characteristics of the intensity of a proliferative reaction. For example it was by estimating the variable $L_S(T)$ that age-related changes in the intensity of proliferative response of the liver to partial hepatectomy were investigated [31].

For quantitative studies of the process of cells' repeated transition to DNA synthesis in SISP use the equality

$$H_S(t) = q_S(t) - L_S(t)\ .$$

Hence, with allowance for (69) we have

$$H_S(t) = q_S(t) - [I_S^c(t) - 1]\exp\left\{\frac{1}{\bar{\tau}_M}\int_0^t I_M(t')dt'\right\} - 1\ .$$

The existence of a fraction of cells repeatedly engaged in DNA synthesis is characterized by the condition

$$q_S(t) - 1 > [I_S^c(t) - 1]\ \exp\left\{\frac{1}{\bar{\tau}_M}\int_0^t I_M(t')dt'\right\},\ t > t_1,$$

from which follows a simple procedure for determining the moment

t_1 when cells appear in the S-phase that resynthesize DNA. The fraction of cells re-entering into the S-phase by the moment t in relation to cells that by that moment have once synthesized DNA is characterized by the index

$$R_S(t) = \frac{q_S(t)-L_S(t)}{L_S(t)} .$$

Noteworthy is the special case of $R_S(t_0)=1$ or

$$q_S(t_0) = 2\left\{1+(I_S^c(t_0)-1)\ \exp\left[\frac{1}{\bar{\tau}_M}\int_0^{t_0} I_M(t')dt'\right]\right\} \tag{72}$$

which formally implies that by the moment t_0 all the cells involved in proliferation could have synthesized DNA at least twice. It should be noted that such assertion may be valid only on the average, since during the time t_0 some of the cells may synthesize DNA more than twice. From equality (72) it follows that $q_S(t_0) \leq 2$.

Kinetic analysis of SISP also envisage investigation of variations in the size of the whole cell population. To estimate increase in the initial total number of cells in the system under study resulting from normal mitotic division (provided no cell death occurs) use may be made of the obvious formula

$$A(t) = \frac{n_{\Sigma(t)}}{n_{\Sigma(0)}} = \exp\left\{\frac{1}{\bar{\tau}_M}\int_0^t I_M(t')dt'\right\} . \tag{73}$$

Obtainable from (73) is a useful estimate for mean duration of mitosis

$$\bar{\tau}_M \geq \frac{\int_0^T I_M(t')dt'}{\ln\ [1+q_S(T)]} . \tag{74}$$

As regards the mean duration of the S-phase it may be well to recall that if $\tilde{I}_S(t) \in L_1(0,\infty)$, then

$$\bar{\tau}_S = \frac{\int_0^\infty \tilde{I}_S(t)dt}{\lim_{t\to\infty} \{q_S(t)\}} . \qquad (75)$$

The last relation follows from the principal renewal theorem,because Γ-distribution is the function of absolutely continuous type (see reference [18] from the previous chapter for definition).

The accuracy of constructing the q-index was investigated by simulation techniques [42,44].

3.6. Examples of Constructing Transient Processes for Particular States of Cell Kinetics

From discussion in 3.4 it follows that for constructing a $v(t,t_0)$ (or $\tilde{v}(t,t_0)$) process it is necessary to define initial cell-age distribution density $w(a,t_0)$ so as to use the expression

$$v(t,t_0) = n(t_0)\int_0^\infty \frac{[1-F(t-t_0+a)]}{[1-F(a)]} w(a,t_0)da . \qquad (76)$$

It will be remembered that only stochastic process expectations are involved in the examination of the problem. It is clear that for defining the age distribution $w(a,t_0)$ in most cases it is necessary to have information on variations in the rate of cell entry into the given phase $k_1(t)$ occurring prior to the moment t. Assuming that the distribution function $F(x)$ belongs to the class of finite functions, i.e. is bounded and equal to zero outside the interval $[0,\tau_{max}]$, for calculating the transient process $v(t,t_0)$ it is sufficient to define $k_1(x)$ on the interval $[t_0-\tau_{max}, t_0]$. Indeed, in that case

$$v(t,t_0) = \int_{t_0-\tau_{max}}^{t_0} k_1(x)[1-F(t-x)]dx.$$

Let us consider some situations for which the system's prehistory is defined by shaping conditions fully determining the state of cell proliferation processes before the moment t_0. Assume at first that by the moment t_0 all the cells in a given phase of the mitotic cycle have been collected at the zero age in respect to the phase. At the moment t_0 the effect of the synchronizing agent is eliminated and the cells begin normal development in the mitotic cycle. Then from (76) it follows that the number of cells remaining in a given phase at the moment $t>t_0$ is equal to

$$v(t,t_0) = n(t_0)[1-F(t-t_0)] ,$$

i.e. we have an expression for $v(t,t_0)$ fully agreeing with the formula which describes the expected number of cells in the first generation of a branching process (Section 2.2.). If cells are synchronized at some other finite age $a_0>0$ of the cycle phase under consideration, the transient process $v(t,t_0)$ is described by the expression

$$v(t,t_0)=\frac{n(t_0)[1-F(t-t_0+a_0)]}{[1-F(a_0)]}.$$

Assume now that a cell population before the moment t_0 was in a strict-sense steady state which may be characterized by the conditions: $n(t)$=const, $n_\Sigma(t)$=const and $k_1=n/\bar{\tau}$ for each of the cell cycle phase. From these conditions follows the formula for the initial density of age distribution in a given phase

$$w(a,t_0) = \frac{1}{\bar{\tau}} [1-F(a)] . \qquad (77)$$

The mean age $\bar{a}$ is equal to: $\bar{a}=m_2/2\bar{\tau}$ if $\bar{\tau}$ and $m_2=\int_0^\infty x^2 dF(x)$

are finite. Substituting (77) in (76) we have

$$v(t,t_0) = \frac{n(t_0)}{\bar{\tau}} \int_0^{\infty} [1-F(t-t_0+a)]da =$$

$$\frac{n(t_0)}{\bar{\tau}} \left\{ \bar{\tau} - \int_0^{t-t_0} [1-F(x)]dx \right\} .$$

Hence, in that case for a transient process the formula

$$v(t,t_0) = n(t_0) \left\{ 1 - \frac{1}{\bar{\tau}} \int_0^{t-t_0} [1-F(x)]dx \right\} \tag{78}$$

is valid. Introducing the designation $A(t)=v(t,t_0)/n(t_0)$, rewrite (78) in the form

$$A(t) = 1 - \frac{1}{\bar{\tau}} \int_0^{t-t_0} [1-F(x)]dx . \tag{79}$$

Apparently, the function $A(t)$ may be put into the form

$$A(t) = \tilde{v}(t,t_0)/I(t_0) = \tilde{v}(t,t_0)/\tilde{v}(t_0,t_0) .$$

In the next subsection the function $A(t)$ will be used in a more detailed study of the process of blocking cells in the mitotic cycle.

If against the background of a strict-sense steady state of a population attained by the moment $t_0=0$ stimulation of cell proliferation is effected at the moment $t_1>t_0$, the $q(t)$ index may be constructed by calculating with formula (60) after substituting in it the difference of functions $\tilde{I}(t)-\tilde{v}(t,0)$. Represented schematically in Fig.3 is the behaviour of the principal kinetic indices $\tilde{I}(t)$, $\tilde{v}(t,0)$ and $\tilde{I}(t)-\tilde{v}(t,0)$ at the hypothetical transition of a population from a steady state to that of stimulated proliferation. Thus, it is evident that the application of the q-index may be extended to SISP with a high initial level of cell proliferation. Fig.3 demonstrates how experimental possibilities are taken advantage of for constructing the function q(t) with a perturbation of cell kinetics in a steady state.

Another state of practical interest is the diurnal rhythm of cell proliferation processes. A mathematical model of that state (see Section 3.3) is based upon representation of the function I(t) with a trigonometric polynomial on the assumption that $n_\Sigma(t)$=const. The functions n(t) and $k_1(t)$ appear in the complex form as

$$n(t) = \sum_{k=0}^{\nu} C_k e^{i\omega_k t} \quad ; \quad k_1(t) = \sum_{k=0}^{\nu} B_k e^{i\omega_k t} \; .$$

Defining the frequency characteristic

$$\Phi(i\omega_k) = \int_0^\infty e^{i\omega_k t}[1-F(t)]dt \; ,$$

the complex amplitudes B_k and C_k may be connected by the relation $C_k=B_k\ \Phi(i\omega_k)$. Reverting to formula (76) we obtain the explicit form of the function $v(t,t_0)$ by means of the following elementary manipulations

$$v(t,t_0) = \int_{-\infty}^{0} \sum_{k=0}^{\nu} B_k e^{i\omega_k\tau}[1-F(t-\tau)]d\tau = \tag{80}$$

$$\sum_{k=0}^{\nu} B_k \Phi(i\omega_k) e^{i\omega_k t}\left\{1 - \frac{1}{\Phi(i\omega_k)} \int_0^{t-t_0} e^{-i\omega_k x}[1-F(x)]dx\right\} =$$

$$\sum_{k=0}^{\nu} C_k e^{i\omega_k t}\left\{1 - \frac{1}{\Phi(i\omega_k)} \int_0^{t-t_0} e^{-i\omega_k x}[1-F(x)]dx\right\} .$$

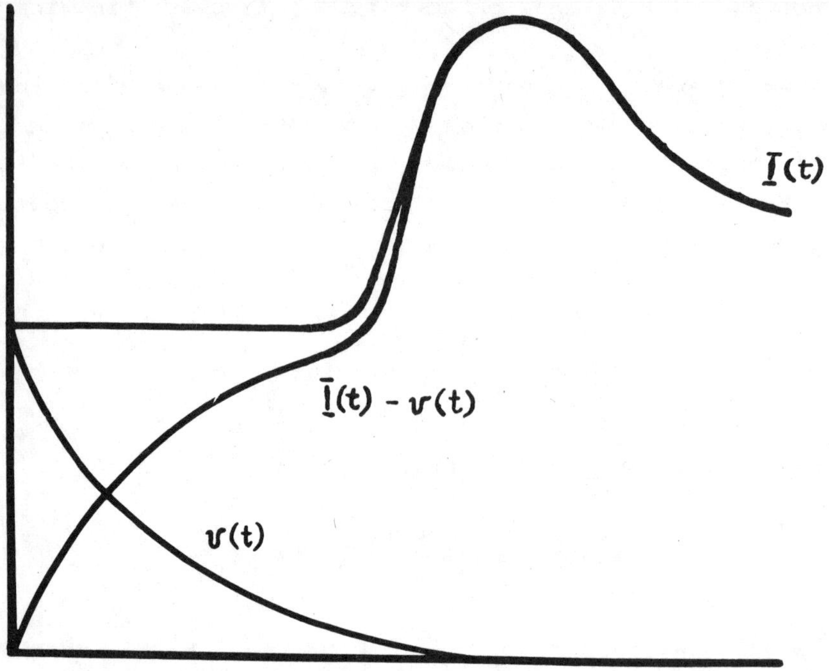

Figure 3. The behaviour of cell kinetic indices in the case of stimulation of proliferative processes against the background of a strict-sense steady state of a given phase of the mitotic cycle.

Thus, the transient process $v(t,t_0)$ in the case under consideration may also be represented in the form of a trigonometric polynomial but now with time varying coefficients. The condition of the constant number of cells in a system makes it possible to substitute in calculations the function $I(t)$ for the function $\tilde{I}(t)$. Stimulation of cell proliferation in a system which before the moment of stimulation was in a state of diurnal rhythm may be interpreted in the same way as stimulation in a steady state, except that the definition of the transient process $v(t,t_0)$ is changed.

From the results outlined in Section 3.2 it follows that in the case of a steady exponential state of a population the cell-age distribution density does not depend upon the current time t and is of the form

$$w(a,t) = \frac{e^{-pa}[1-F(a)]}{\int_0^\infty e^{-pa}[1-F(a)]da} .$$

Substituting this expression in (76) and carrying out the change of variables: $u=t-t_0+a$ for a transient process considered from the moment t_0, we have

$$v(t,t_0) = n(t_0)e^{p(t-t_0)}\left\{1 - \frac{\int_0^{t-t_0} e^{-pu}[1-F(u)]du}{\int_0^\infty e^{-pu}[1-F(u)]du}\right\} ,$$

which, indeed, may be regarded as a special case of relation (80).

The question of constructing transient processes for different states of cell proliferation kinetics will be dealt with once again in the next chapter in connection with examining labelled mitosis curves.

3.7. Analysis of the Process of Cell Blocking in the Mitotic Cycle

It is generally known that with a short-term ionizing irradiation of renewing tissues the mitotic index reduces to zero with subsequent recovery after a certain period when no mitoses occur in the tissue. In analysing the mitotic index drop curve (the first portion of a radiation block curve) it is usually believed that irradiation effect is caused by the block $G_2 \rightarrow M$, however with no effect on cells which at the moment are in a phase of mitosis and which normally complete mitotic division. The validity of that assumption is supported by a concurrent shift of radiation block curves occurring with variation of irradiation dosage within a fairly wide range [32]. This may be interpreted as a displacement of the blocking point towards the start of mitosis with increase in irradiation dose, without changes in the temporal parameters of the process of mitotic cell division.

Ionizing radiation also causes the blocking of DNA synthesis [22,42]. In general, many cytostatic effects result in the appearance of blocks at different points of the mitotic cycle of cells.

In order to construct a mathematical formalization of the process of cell blocking in the cycle let us revert to the principal equation of semistochastic cell kinetics in form (45). Let us consider some phase of the life cycle of cells and assume that at the moment $t=0$ at the beginning of the phase a block appears which stops completely the flow of new cells into the phase. Setting in equation (45) $k_1(t)=0$ for all $t>0$ we see that the number of cells in the given phase in the presence of a block varies in accordance with the dynamics of the transient process $v(t,0)$. In other words, denoting with $n(t,0)$ the mean number of cells remaining in the phase up to the moment $t>0$ we can write

$$n(t,0) = v(t,0) , \tag{81}$$

or

$$n(t,0) = n(0) \int_0^{\infty} \frac{[1-F(t+\tau)]}{[1-F(\tau)]} w(\tau,0)d\tau . \tag{82}$$

Hence, as a problem, mathematical description of complete blocking of cell entry into a given mitotic cycle phase is equivalent to the problem of constructing a transient process for the phase which was discussed in the preceding subsection. Similarly, if a block originates at some moment $t_0>0$, then $n(t,t_0)=v(t,t_0)$. It should also be noted that the function $A(t)=1-n(t,0)/n(0)$ is the <u>distribution function of the residual life-time of cells</u> in the mitotic cycle phase under review.

Assume that prior to the moment of irradiation t_0 the cell system was in a strict-sense steady state, i.e. $k_{1M}=n_M/\bar{\tau}_M$ for $t \in (-\infty,t_0)$. Setting $t_0=0$ and dividing both sides of expression

(82) by $n(0)$ for the function

$$\bar{A}_M(t) = \frac{n_M(t,0)}{n_M(0)} = \frac{I_M(t,0)}{I_M(0)}$$

we obtain

$$\bar{A}_M(t) = 1 - \frac{1}{\bar{\tau}_M} \int_0^t [1-F_M(x)]dx \qquad (83)$$

or

$$A_M(t) = 1 - \bar{A}_M(t) = \frac{1}{\bar{\tau}_M} \int_0^t [1-F_M(x)]dx \ . \qquad (84)$$

It is now quite clear that the function $A(t)$ is a stationary distribution function for the residual life-time of cells in mitosis which has received adequate study for stationary renewal processes [8,14]. Specifically, with an error not exceeding $\frac{1}{2}\left(\frac{t}{\bar{\tau}_M}\right)^2$ the function $\bar{A}(t)$ can be described with the approximate formula

$$\bar{A}(t) \simeq 1 - \frac{t}{\bar{\tau}_M} \ . \qquad (85)$$

Besides, the mean value of the residual life-time of cells in mitosis $\bar{\tau}_0$ is numerically equal to the area under the curve $\bar{A}(t)$ and is connected with the parameters $\bar{\tau}_M$ and σ_M by the following relation

$$\bar{\tau}_0 = \int_0^\infty \bar{A}(t)dt = \frac{\sigma_M^2 + \bar{\tau}_M^2}{2\bar{\tau}_M} \ . \qquad (86)$$

Formulas (85) and (86) provide a graphical way of obtaining estimates of the parameters $\bar{\tau}_M$ and σ_M represented schematically in Fig.4. Such estimates may be used as initial

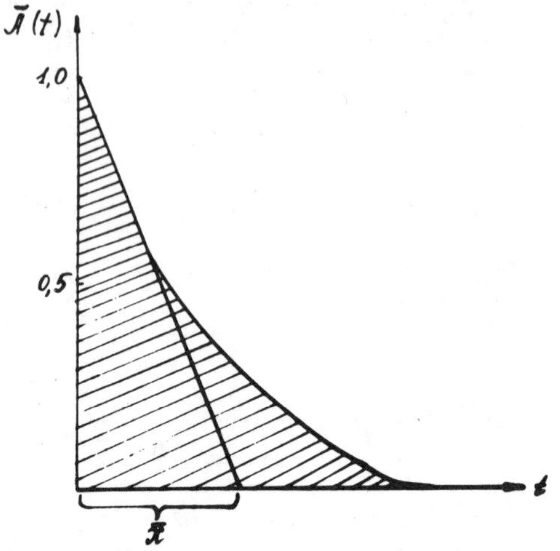

Figure 4. A graphical estimation of the temporal parameters of a cycle phase with a blocked cell entry into the phase.

values in the corresponding algorithm of parameter optimization.

With graphically obtained $\bar{\tau}_M$ and σ_M estimates it is possible to test the model for the radiation block of mitotic activity. Reference [40] describes such a test using approximation of $F_M(x)$ with the Γ-distribution function. In that case (83) has the form

$$\bar{A}(t)=1 - \frac{1}{\bar{\tau}_M} \int_0^t [1-\gamma(\alpha,\beta\tau)]d\tau \quad , \tag{87}$$

where $\gamma(\alpha,\beta\tau)$ is the incomplete Γ-function and $\alpha = \bar{\tau}_M^2/\sigma_M^2$, $\beta = \bar{\tau}_M/\sigma_M^2$.

Calculation with formula (87) of the theoretical A(t) curve at $\bar{\tau}_M$=0.5 hr and σ_M=0,25 hr(the values were determined graphically from experimental data) has shown a good agreement with experimental data [7] on the drop of the mitotic index in the mouse duodenal epithelium after acute gamma-irradiation with a dose of 4 Gy. The reason for considering the calculation as a validation of the mathematical model lies in the fact that the parameters $\bar{\tau}_M$ and σ_M in that case are not free, being directly determined (predicted) by the structure of the model concerned.

Thus, by means of formulas (85) and (86) estimates may be obtained of the first two moments of the F(x) distribution. The curve for a drop of cell fraction in a given cycle phase with a block occurring at the start of the phase also contains information on the higher moments of the probability distribution F(x) . For example, the third μ_3 and fourth μ_4 initial moments are calculated from the formulas

$$\mu_3=6\bar{\tau}\int_0^{\infty} t\bar{A}(t)dt \ ; \quad \mu_4=12\bar{\tau}\int_0^{\infty} t^2\bar{A}(t)dt,$$

which, obviously, prevail with fairly mild restrictions on the class of F(x) distributions. In general, if there exists the limit

$$\lim_{t\to\infty} t^k[1-F(t)]=0$$

it is possible for the k-th moment of distribution F(x) to write

$$\mu_k=\bar{\tau}k(k-1)\int_0^{\infty} t^{k-2}A(t)dt \ . \qquad (88)$$

Now let us include in the consideration of cell blocking in the cycle the possible death of cells after the moment of the initiation of a block. We shall consider, as before, that prior to the initiation of a block the cell system was in a strict sense steady state assuming besides that cell death exerts no appreciable effect on the expected total size of the whole cell system n_Σ . We shall further assume that a cell may either die, losing signs of belonging to a given phase (type 1 failure) or else normally complete passage through the given phase (type 2 failure). Let the potential time of cell life in a given phase prior to the onset of types 1 and 2 failures be equal to X_1 and X_2, respectively. The random variables X_1 and X_2 are considered to be independent and possessing different distribution functions $\mathcal{L}_1(x)$ and $\mathcal{L}_2(x)$ (with the means $\bar{x}_1$ and $\bar{x}_2$ and variances $\sigma^2_{X_1}$ and $\sigma^2_{X_2}$). This scheme of two competing independent risks is sometimes referred to as the "double risk scheme". The same scheme was used by Nedelman et.al. (reference [16] in Chapter II) in their work aimed at gaining inference (unobserved parameters estimation) from growth patterns of mast-cell colonies on the basis of age-dependent multitype branching process model.

The density of distribution of the observed phase duration, i.e. of the variable $X=\min\{X_1,X_2\}$ is defined by the expression

$$f(x) = \ell_1(x)[1 - \mathcal{L}_2(x)] + \ell_2(x)[1 - \mathcal{L}_1(x)],$$

where $\ell_i(x)$ are the densities of distribution of the variables X_i, i=1,2. Hence, the reliability (survival) function for X is equal to

$$\bar{F}(x) = 1-F(x) = [1-\mathcal{L}_1(x)]+[1-\mathcal{L}_2(x)]-[1-\mathcal{L}_1(x)\mathcal{L}_2(x)] , \quad (89)$$

and, consequently, for the mean value we have

$$\bar{\tau} = \int_0^\infty [1-F(x)]dx = \bar{x}_1 + \bar{x}_2 - \int_0^\infty [1-\mathcal{L}_1(x)\mathcal{L}_2(x)]dx . \quad (90)$$

Expression (84) holds true if the function $\bar{F}(x)$ and the number $\bar{\tau}$ are defined in accordance with (89) and (90). Consider a special case when $\ell_1(x)=\lambda e^{-\lambda x}$ (a purely random cell death) and $\ell_2(x)$ is the Γ-distribution density with the parameters $\alpha=\bar{x}_2^2/\sigma^2_{X_2}$ and $\beta=\bar{x}/\sigma^2_{X_2}$. Then for the function A(t) we have the following formula

$$A(t) = \frac{\lambda(\beta+\lambda)^\alpha}{(\beta+\lambda)^\alpha - \beta^\alpha} \int_0^t e^{-\lambda x}[1-\mathcal{L}_2(x)]dx , \quad (91)$$

where

$$\mathcal{L}_2(x) = \frac{\beta^\alpha}{\Gamma(\alpha)} \int_0^x z^{\alpha-1} e^{-\beta z} dz \ .$$

In the absence of cell death, i.e. with $\lambda \to 0$, we have

$$\lim_{\lambda\to 0} A(t) = \frac{1}{\bar{x}_2} \int_0^t [1-\mathcal{L}_2(x)] dx \ .$$

Fig.5 illustrates the effect of cell death rate on the form of the $\bar{A}(t)$ curve calculated with formula (91). It is evident from Fig.5 (and directly from formula (91)) that there is interaction between the parameters λ and $\bar{x}_2$ which complicates considerably their estimation. It is clear that only generalized parameters of the life-time X, i.e. $\bar{\tau}$ and σ are estimated by means of the graphical method. Fig.5 is a good illustration of nonidentifiability of the characteristics of marginal distributions $\mathcal{L}_1(x)$ and $\mathcal{L}_2(x)$ by observations of the residual life-time distribution $A(t)$ even within the framework of the scheme of independent competing risks. The problem of the nonidentifiability of marginal distributions in the general scheme of dependent competing risks is dealt with in references [21,27, 37] under the circumstances when rough survival probabilities $Q_i(t)=\mathbb{P}\{X_i>t, \bigcap_{i\neq j} X_i<X_j\}$, $i=1,2,\ldots,k$, are accessible to observation.

Let us try to modify somewhat the scheme of cell death. Assume that with the probability p the life-time of a cell in a given phase is the variable X_1 and with the probability $1-p$ it is X_2. As before, the variables X_1 and X_2 have the probability distributions $\mathcal{L}_1(x)$ and $\mathcal{L}_2(x)$. Obviously, this scheme (with a single risk) presumes, in fact, the presence (following the initiation of a block) of two different cell populations corresponding to the cycle phase under study, viz., one of dying cells and the other of those normally completing their evolution in the phase. In that case the phase duration distribution function appears as a simple two-component mixture

$$F(x) = p\mathcal{L}_1(x) + (1-p)\mathcal{L}_2(x) , \tag{92}$$

and the mean value has the form

$$\bar{\tau} = p\bar{x}_1 + (1-p)\bar{x}_2 .$$

This scheme was already discussed in the preceding chapter in connection with generalization by Jagers of the branching process model and obtaining the maximum likelihood estimator of the cell death factor [20] in the exponential steady state of a cell population.

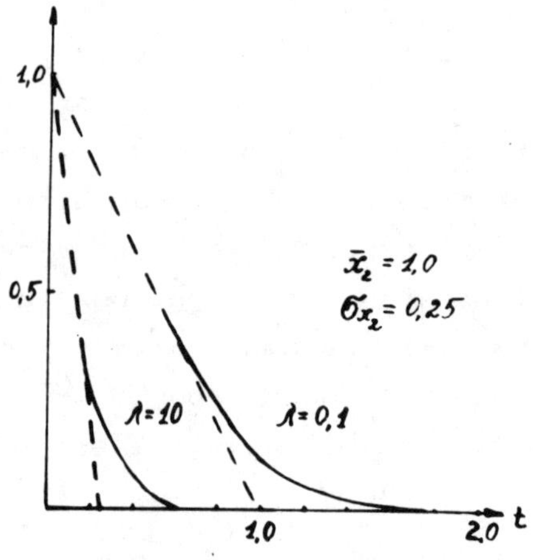

Figure 5. The effect of cell death (the "double risk" death scheme) on the form of the curve for blocking cell entry into a cycle phase. λ – cell death rate.

In the case under review the formula valid for $A(t)$ is

$$A(t) = \frac{p\int_0^t [1-\mathcal{L}_1(x)]dx + (1-p)\int_0^t [1-\mathcal{L}_2(x)]dx}{p\bar{x}_1 + (1-p)\bar{x}_2}$$

For the initial moments of the variables X_1 and X_2 we may write

$$\mu_j^X = p\,\mu_j^{X_1} + (1-p)\,\mu_j^{X_2} ,$$

and the moments μ_j^X may be calculated according to formula (88). If the variable X_1 is distributed in conformity with the exponential law while X_2 has the Γ-distribution, the following set of non-linear equations is formed for the sought - for parameters

$$\mu_k^X = p\,\frac{k!}{\lambda^k} + (1-p)\,\frac{\alpha(\alpha+1)\ldots(\alpha+k-1)}{\beta^k} ,k=1,2,\ldots,m .$$

Hence, the death scheme with a single risk makes it possible to supplement the principal equation for $A(t)$ with a system of connections superimposed on the distribution moments involved in mixture (92). Such treatment also appears essential in analyzing curves for in vitro maturation of reticulocytes the principles of which were elaborated by Mosyagina [25]. It should be noted that the results obtained by Mosyagina agree with the more general model presented here for the case of a strict-sense steady state and no cell death. As for the situation when a block occurs before cells enter into the cycle phase in question, model experiments on a computer have shown that the interval between $t=0$ and the moment the number of cells begins to drop is a satisfactory estimate of the mean duration of the portion of the cycle that separates the blocking point from the start of the phase concerned (provided the population was in a steady state before the block).

REFERENCES

1. Bartlett,M.S. Distributions associated with cell populations, Biometrika, 56, 391-400, 1969.
2. Bartlett,M.S. Age distributions, Biometrics, 26, 377-385, 1970.
3. Bellman,R. and Cook, K.L. Differential-difference equations, Academic Press, New York-London, 1963.
4. Brown,J.M. and Berry,R.J. The relationship between diurnal variation of the number of cells in mitosis and the number of cells synthesizing DNA in the epithelium of the hamster cheek pouch, Cell Tiss.Kinet., 1, 23-33, 1968.
5. Bullough,W.S. and Laurence, E.B. The diurnal cycle in epidermal mitotic duration and its relation to chalone and adrenaline, Exper.Cell Res., 43, 343-350, 1966.
6. Cameron, I.L. and Greulich, R.C. Evidence for an essentially constant duration of DNA synthesis in renewing epithelia of the adult mouse, J.Cell Biol., 18, 31-40, 1963.
7. Chumak, M.G. The effect of radiation on cell division in tissues differing in radiosensitivity, In: Repair Processes Following Radiation Injury, Moscow, 47-52,1964 (In Russian).
8. Cox, D.R. Renewal theory, Methuen and Co LTD,London,John Wiley and Sons Inc., N.Y., 1962.
9. Ellem, K.A.O. and Mironescu, S. The mechanism of regulation of fibroblastic cell replication. I. Properties of the system, J.Cell.Physiol., 79, 389-406, 1972.
10. Fabrikant, J.I. The kinetics of cellular proliferation in regenerating liver,J.Cell Biol.,36, 551-565,1968.
11. Feller, W. On the integral equation of renewal theory, Ann.Math.Stat., 12, 243-267, 1941.
12. Feller, W. An introduction to pobability theory and its applications, vol.2, John Wiley and Sons, Inc., N.Y.-London-Sydney-Toronto, 1971.
13. Von Foerster H., Some remarks on changing populations, In: The Kinetics of Cellular Proliferation, Stohleman F.(ed.), N.Y., 382-407, 1959.
14. Gnedenko, B.V., Belyaev, Yu.K. and Solovyev, A.D. Mathematical methods of the reliability theory, Nauka, Moscow, 1965(In Russian).
15. Grobe, D.D. Auerbach,H. and Brues, A.M. Diurnal variation in the labeling index of mouse epidermis. A double isotope autoradiographic demonstration of changing flow rates, Cell Tiss.Kinet., 3, 363-373, 1970.
16. Harris, T.E. The theory of branching processes, Springer-Verlag, Berlin, 1963.
17. Hartmann, N.R. and Moller, U. A compartment theory in the kinetics including considerations on circadian variations, In:Biomathematics and Cell Kinetics, Elsevier/North-Holland, Amsterdam, 223-251, 1978.
18. Himmelblau, D.M. Process analysis by statistical methods, John Wiley and Sons, Inc., New York-London-Sydney-Toronto, 1970.
19. Hurgin, Yu.J. and Yakovlev, V.P. Finite functions in physics and technology, Nauka, Moscow, 1971 (In Russian).
20. Jagers, P. Branching processes with biological aplications, John Willey and Sons, N.Y., Sydney, Toronto, 1975.

21.Kalbfleisch, J.D., and Prentice, R.L. The statistical analysis of failure time data, Willey, N.Y., 1980.

22.Lesher, G. and Lesher, S. Effect of single dose whole-body ^{60}Co γ-radiaton on number of cells in DNA synthesis and mitosis in mouse duodenum epithelium, Rad.Res., 43,429-438, 1970.

23.MacDonald, P.D.M. Statistical inference from the fraction labelled mitoses curve,Biometrika,57,489-503,1970.

24.Martinez,H.M. On the derivation of mean growth equation for cell cultures,Bull.Math.Biophys.,28,411-416,1966.

25.Mosyagina, E.N. Erythrocytic equilibrium under normal and pathologic conditions, Meditzina, Moscow, 1962 (In Russian).

26.Pilgrim, C., Erb, W. and Mauer, W. Diurnal fluctuations in the number of DNA synthesizing nuclei in various mouse tissues, Nature, 199, 863,1963.

27.Prentice,R.L.,Kalbfleisch, J.D., Peterson, A.V., Flournoy, N., Farewell, V.T. and Breslow, N.E. The analysis of failure times in the presence of competing risks, Biometrics, 34, 541-554,1978.

28.Rao, P.N. and Engelberg, J. Mitotic duration and its variability in relation to temperature in HeLa cells, Exper.Cell Res., 52, 198-208, 1968.

29.Rosenwasser, E.N. Periodically non-stationary control systems, Nauka,Moscow, 1973 (In Russian).

30.Serebrennikov, M.G. and Pervozvansky, A.A. Detection of latent periodicities,Nauka, Moscow, 1965 (In Russian).

31.Stöcker,E., Schultze,B., Heine, W.D. and Liebcher,H. Wachstum und Regeneration in parenchymatösen Organen der Ratte, Z.Zellforsch., 125, 306-331, 1972.

32.Strzhizhovsky, A.D. On the effect of ionizing radiation on the duration of mitosis in mouse corneal epithelium, Radiobiologiya, 4,476-482,1964 (In Russian).

33.Tikhonov, A.N. and Arsenin, V.Y. Solutions of ill-posed problems, Wiley/Winston,London, 1977.

34.Titchmarsh, E.C. An introduction to Fourie integrals theory, OGIZ, Moscow-Leningrad, 1948 (In Russian translation).

35.Tricomi, F.G. Integral equations, Interscience Publisher, New York, London, 1957.

36.Trucco, E. Mathematical models for cellular systems. The Von Foerster equation. Part I, Bull. Math.Biophys., 27, 285-304, 1965.

37.Tsiatis, A. A nonidentifiability aspect of the problem of competing risks, Proc.Nat.Acad.Sci. USA,72,20-22,1975.

38.Tvermur, E.M.F. Circadian rhythms in epidermal mitotic activity.Diurnal variations of the mitotic index,the mitotic rate and the mitotic duration, Virch.Arch.Abt.B.,Zellpath., 2, 318-328, 1969.

39.Yakovlev, A.Yu. On certain possibilities of investigating a system of steady-state tissue regulation, Cytology, 13, 1417-1425, 1971(In Russian).

40.Yakovlev, A.Yu. On simulation of the radiation block of mitotic activity, Cytology, 15, 616-619, 1973(In Russian).

41.Yakovlev, A.Yu. Diurnal proliferation rhythm and mitotic cycle phase parameters, Cytology, 15, 473-475, 1973(In Russian).

42.Yakovlev, A.Yu. and Zorin,A.V. Computer simulation in cell radiobiology, Springer-Verlag, Berlin, Heidelberg, New York, 1988.

43. Yakovlev, A.Yu., Zorin, A.V. and Isanin, N.A. The kinetic analysis of induced cell proliferation, J.Theor. Biol., 64, 1-25, 1977.

44. Zorin, A.V. and Yakovlev, A.Yu. The properties of cell kinetic indicators. A computer simulation study, Biom.J., 3,347-362, 1986.

IV. THE FRACTION LABELLED MITOSES CURVE IN DIFFERENT STATES OF CELL PROLIFERATION KINETICS

4.1. Introduction

The analysis of the fraction labelled mitoses curve (FLM) isone of the most frequently used methods for estimating indirectly the numerical characteristics (mean and variance) of the lengths of the separate phases in the mitotic cycle. Nearly all experimental data on cell population kinetics whether "in vitro" or "in vivo" are interpreted by analysing the structure of FLMs. In the majority of such studies graphical methods are used for the estimation of the mitotic cycle and its phases lengths; these are not based on modern dynamic theory of cell systems. In order to obtain more sound methods it is necessary to construct a mathematical model for FLM and an appropriate procedure for non-linear estimation of its parameters.The basic principle for the indirect estimation of the mitotic cycle temporal parameters consists, therefore, of finding a set of their values which maximize, in some sense, the agreement between the model and the experimentally found FLM.

The methods for describing the FLM curve mathematically which have been developed up to present [see for review: 11,20,35] are usually applicable only to cell populations with stationary cell age (in respect to the starting point of the mitotic cycle) distribution - such populations either grow exponentially or are in the steady-state (in strict sense) of growth. This fact will be proved below. The works of MacDonald [32] and Jagers [26] contain the fullest theoretical solution for the FLM problem under conditions of exponential state of cell population kinetics. Using the approach proposed there, several analyses of experimental data have been carried out, which led to a series of new results concerning the behaviour of real cell populations [8,9,10,15, 19,27,39,43,44,45] .In experimental studies, however, more complex situations can be tackled by analysing the FLM curve, e.g. diurnal variation in cell proliferation processes, transient cell kinetics in systems with induced DNA synthesis or in synchronized populations. That is why the mathematical and computer simulation methods in the analysis of FLM deserve further development.

4.2. "Flux-expectations" Concept and the Fraction Labelled Mitoses Curve

Let us return to the basic integral equation of semistochastic cell kinetics introduced in the previous chapter

$$\mathbb{E}\{N(t)\} = \int_0^t \mathbb{E}\{K_1(\tau)\} \, [1-F(t-\tau)] \, d\tau + v(t), \tag{1}$$

where

$$v(t) = \bar{N}(0) \int_0^\infty \frac{1-F(t+\tau)}{1-F(\tau)} \, w(\tau,0) \, d\tau$$

and $\bar{N}(0)$ is the initial expected number of cells in the population.

The function $v(t)$ reflects the influence the initial state of the population has on the cell kinetics, i.e., it can be thought of as the process of change in the numbers of those cells which at time $t=0$ comprised the whole population (the transient process). The initial condition (and, therefore, the dynamic history of the cell population) are, in this case, given by the initial mean number of cells in the cycle phase $\bar{N}(0)$ and the initial age distribution $w(a,0)$. In the problem of constructing FLM curves, the function $v(t)$ is interpreted as follows. At the moment $t=0$, let those cells synthesising DNA also incorporate ^{3}H-thymidine, so that the time necessary for including the labelled precursor is assumed to be equal to zero. Then $\bar{N}_S(t)$ is the mean (expected) number of cells which have incorporated the impulse label and $v_S(t)$ is the mean number of cells which were initially labelled and which have not left phase S by time t. By ignoring the possibility of repeated DNA synthesis by previously labelled cells, for any time $t>0$ one can compute the stream of labelled cells from phase S into phase G_2:

$$r^{*}_{G_2}(t) = \int_0^t \mathbb{E}\,\{K^{*}_{1,G_2}(\tau)\}d\tau = \int_0^t \mathbb{E}\{K^{*}_{2,S}(\tau)\}d\tau = \bar{N}_S(0) - \bar{v}_S(t)\ . \quad (2)$$

Here and afterwards symbols with asterisks denote parameters of labelled cells.

Thus equation (2) is the starting point in the study of the first FLM wave, if one takes the function $f(x)$ (or $F(x)$) as the basic dynamic characteristic of each phase in the mitotic cycle.

The majority of authors do not use equation (1), preferring a different way of constructing the theoretical FLM curve - the so called "method of flux-expectations". The main feature of this method is the use of the probability density of the age of the cell at the moment when it leaves the phase, $h\ (a,t)$ .In other words, $h(a,t)$ is defined for cells picked at the end of each phase in the mitotic cycle. Here the time t is considered as a continuous parameter. Returning to the previously introduced notion of the age of a cell in phase (A) and the duration of phase (X) we may write

$$h(a\ ,t) = \lim_{\Delta a \to 0} \frac{\mathbb{P}_t\{a < A \le a+\Delta a \mid A=X\}}{\Delta a}\ ,$$

where the subscript t on $\mathbb{P}$ means that a population is examined of fixed time t.

Recall the definition of the failure rate function $\lambda(a,t)$ and represent it in the following form

$$\lambda(a,t) = \lim_{\Delta a \to 0} \frac{\mathbb{P}_t\{a < X \le a+\Delta a \mid a < A \le a+\Delta a\}}{\Delta a}\ . \qquad (3)$$

In view of (3) it is clear that

$$\lambda(a,t) = \lim_{\Delta a \to 0} \frac{\mathbb{P}_t\{a<X\leq\Delta a\ ,\ a<A\leq a+\Delta a\}\ \Delta a}{\Delta a\ \Delta a\ \mathbb{P}_t\{a<A\leq a+\Delta a\}} =$$

$$\frac{1}{w(a,t)} \lim_{\Delta a \to 0} \frac{\mathbb{P}_t\{a<X\leq a+\Delta a\ ,\ a<A\leq a+\Delta a\}}{\Delta a\ \Delta a} = \frac{g(a,t)}{w(a,t)}\ .$$

where $g(a,t)=f_{X,A}(a,a;t)$, and $f_{X,A}(x,a;t)$ is the joint distribution density for the random variables X and A. Hence

$$g(a,t) = \lambda(a,t)w(a,t).$$

On the other hand we may write

$$g(a,t) = \lim_{\Delta a \to 0} \frac{\mathbb{P}_t\{a<A\leq a+\Delta a\ |\ A=X\}\ \mathbb{P}_t\{A=X\}}{\Delta a\ \Delta a} = h(a,t)\int_0^\infty g(u,t)du.$$

Now we have the desired expression for $h(a,t)$:

$$h(a,t) = \frac{\lambda(a,t)\ w(a,t)}{\int_0^\infty \lambda(u,t)\ w(u,t)du}\ .$$

In the particular and most interesting case in which the failure rate λ doesn't depend on t, we obtain

$$h(a,t) = \frac{\lambda(a)\ w(a,t)}{\int_0^\infty \lambda(u)\ w(u,t)du} = \frac{\mathbb{E}\{K_1(t-a)\}\ f(a)}{\int_0^\infty \mathbb{E}\{K_1(t-a)\}\ f(a)da}\ .$$

The denominator in the last expression is equal to $\mathbb{E}\ \{K_2(t)\}$ (see formula (13) from Chapter III); hence the final expression for $h(a,t)$ has the form

$$h(a,t) = \mathbb{E}\ \{K_1(t-a)\}\ f(a)/\mathbb{E}\ \{K_2(t)\}. \qquad (4)$$

Having obtained the function $h(a,t)$, it is possible to write a general expression for the FLM curve. Indeed, the expected fraction of cells which are labelled at time $t=0$ among all cells which, for the first time after labelling entered the phase of mitosis at time $t>o$, can be expressed as

$$l_0^*(t) = \int_0^t h_{G_2}(a,t)\, da - \int_0^t h_{S+G_2}(a,t)\, da =$$

$$\int_0^t \frac{\mathbb{E}\{K_{1,G_2}(t-a)\}f_{G_2}(a)}{\mathbb{E}\{K_{2,G_2}(t)\}}\, da - \int_0^t \frac{\mathbb{E}\{K_{1,S+G_2}(t-a)\}f_{S+G_2}(a)}{\mathbb{E}\{K_{2,S+G_2}(t)\}}\, da, \quad (5)$$

and for the mathematical expectation of the fraction of labelled mitoses FLM (t) we have

$$FLM(t) = \int_0^t l_0^*(t-\tau)\; w_M\,(\tau,t)\; d\tau.$$

This approach to finding FLM (t) turns out to be convenient only in the case when $h(a,t)$ is independent of t and, therefore, simple analytical expressions for $h(a)$ are admissible. In practical applications, the assumption that the mitotic phase has a stationary age distribution is also essential. These conditions are satisfied in a stationary exponential state, when for every phase $\mathbb{E}\,\{K_1(t)\} = Ce^{pt}$ and the age distribution is stationary. Then it is easy to see, that

$$\mathbb{E}\{K_2(t)=\mathbb{E}\{N(t)\}\int_0^\infty \lambda(a)w(a)da=Ce^{pt}\int_0^\infty e^{-px}f(x)dx,$$

and therefore, expression (4) in the case of exponentially growing populations takes the form

$$h(a) = e^{-pa}f(a) \;/\; \int_0^\infty e^{-px}f(x)\; dx. \qquad (6)$$

It is easy to write an inverse relationship too

$$f(a) = e^{pa}h(a) \Big/ \int_0^\infty e^{px}h(x)\,dx. \tag{7}$$

The rigorous derivation of this result, based on the limiting age distribution is given in the work by MacDonald [32]. Under conditions of strict steady state of the population, i.e., $\mathbb{E}\{K_{1,S+G_2}(t)\} = \mathbb{E}\{K_{1,G_2}(t)\} = \mathbb{E}\{K_{2,G_2}(t)\}=$ constant , we arrive at the well-known formula by Barrett [2], which, obviously, can also be obtained by substituting $p=0$ in (6) or (7). Since, in the exponential state, one may assume that both descendant cells formed by a division are returned to the mitotic cycle, it is not difficult, in principle, to consider the possibility of multiple entries of labelled cells into mitosis. Under the additional assumption that the durations of consecutive mitotic cycles (and of their phases) are mutually independent, expression (5) can be generalized as follows:

$$l^*(t)= \int_0^t h_{G_2}(x)dx- \int_0^t h_{G_2} * h_S(x)dx + \sum_{i=1}^{\infty} \int_0^t [h_{G_2} * h_c^{*i}(x)-h_{G_2} * h_S * h_c^{*i}(x)]dx. \tag{8}$$

Here h_c^{*i} is the i-fold convolution of the function

$$h_c(x)=h_{G_1} * h_{G_2} * h_S(x) .$$

If we take Laplace transforms of both sides of equation (8) it simplifies significantly:

$$L^*(p)= \frac{H_{G_2}(p)[1-H_S(p)]}{p\,[1-H_c(p)]} \qquad \text{for} \quad Re(p)>0,$$

and can be used in numerical computations and in the study of the asymptotic properties of the function $l^*(t)$ [41]. Laplace transform in some cases also allows a very simple transition from $f(x)$ to $h(x)$ and vice-versa, changing only the numerical parameters of the distribution [32,41]. Let $f(x)$ be the Γ-distribution density with the form parameter $\alpha = \bar{\tau}^2/\sigma^2$ and the scale parameter $\beta = \tau/\sigma^2$. Then $h(x)$ will also be the Γ-distribution density with the same form parameter $\alpha' = \alpha$ but with the scale parameter $\beta' = \beta + p$. On the contrary, if approximation $h(x)$ with Γ-distribution is used with the parameters α' and β', then $f(x)$, retaining the analytical form, will have the parameters $\alpha = \alpha'$ and $\beta = \beta'/[1-\beta' p]$. In that way one can estimate an error occurring in the formal use of the Barrett method [1] for analyzing the FLM curve in a growing population of cells. The population growth index p is determined from the characteristic equation $\eta\hat{f}(p)=1$, where $\hat{f}(p)$ is the Laplace transform of the density of mitotic cycle duration and η is the generation coefficient.

Formula (4) (hence (6) and (7) as well) may be generalized to the case of the combination of r successive cell cycle phases whose durations are dependent random variables, and cell age is measured from the start of the last (r-th) phase.

Let us consider the simplest variant of combination of two ($r=2$) successive phases 1 and 2 with the durations X_1 and X_2, respectively, with the variables X_1 and X_2 characterized by the joint probability density $f_{1,2}(x_1,x_2)$. The duration of the combined phase 1+2 is $X=X_1+X_2$, and therefore it can be described by the one-dimensional probability density

$$f_{1+2}(x) = \int_0^x f_{1,2}(x_1, x-x_1)\,dx_1 . \tag{9}$$

If the corresponding function $h(a,t)$ is defined for the phase 1+2, formula (4) retains its previous form, i.e

$$h_{1+2}(a,t) = \mathbb{E}\{K_1^{(1)}(t-a)\} f_{1+2}(a) / \mathbb{E}\{K_2^{(2)}(t)\}, \tag{10}$$

where the superscripts indicate the numbers of phases for which the functions $K_1(t)$ and $K_2(t)$ are considered.

Suppose the age of the cell leaving phase 1+2 is measured from the start of phase 2 and find under such conditions the form of the function $h_2(a_2,t)$. The required formula for $h_2(a_2,t)$ may be obtained by virtue of the consideration that

$$h_2(a_2,t)=\int_0^\infty p_t(a_2,a)da, \qquad (11)$$

where $p_t(a_2,a)$ is the joint density of distribution of cell age for phase 2 and for combined phase 1+2 defined for cells just completing phase 2. Let $f(x_2|x)$ be the conditional probability density of phase 2 duration at a given length of phase 1+2, then

$$p_t(a_2,a)=h_{1+2}(a)f(a_2|a) \ . \qquad (12)$$

For $f(x_2|x)$, in turn, we can write

$$f(x_2|x)=f_{1,2}(x-x_2,x_2)/f_{1+2}(x), \qquad (13)$$

and , substituting (10) and (13) in (12), we have

$$p_t(a_2,a)=\mathbb{E}\{K_1^{(1)}(t-a)\}f_{1,2}(a-a_2,a_2)/\mathbb{E}\{K_2^{(2)}(t)\}. \qquad (14)$$

Next, turning to formula (11) we obtain from (14) the following expression

$$h_2(a_2,t)= \int_0^\infty \mathbb{E}\{K_1^{(1)}(t-a)\}f_{1,2}(a-a_2,a_2)da/\mathbb{E}\{K_2^{(2)}(t)\} \ . \qquad (15)$$

Changing the variables: $a = a_1+ a_2$ will put formula (15) into a more instructive form

$$h_2(a_2,t)= \int_0^\infty \mathbb{E}\{K_1^{(1)}(t-a_1-a_2\}f_{1,2}(a_1,a_2)da_1/\mathbb{E}\{K_2^{(2)}(t)\}. \qquad (16)$$

In the case of independent variables X_1 and X_2 formula (16) reduces to formula (4) applied only to phase 2.

Using the relation

$$\mathbb{E}\{K_2^{(2)}(t)\}= \int_0^\infty \mathbb{E}\{K_1^{(1)}(t-a)\}f_{1+2}(a)da \qquad (17)$$

and formula (9) we come to the following final expression

$$h_2(a_2,t)=\frac{\int_0^\infty \mathbb{E}\{K_1^{(1)}(t-a_1-a_2)\}f_{1,2}(a_1,a_2)da_1}{\int_0^\infty\int_0^\infty \mathbb{E}\{K_1^{(1)}(t-a_1-a_2)\}f_{1,2}(a_1,a_2)da_1da_2} . \tag{18}$$

For the case of a combination from any finite number r of successive cycle phases and cell age measured from the start of the r-th phase formula (18) takes the form

$$h_r(a_r,t)=\frac{\int_0^\infty\cdots\int_0^\infty \mathbb{E}\{K_1^{(1)}(t-\sum_{i=1}^{r}a_i)\}f_{1,\ldots,r}(a_1,\ldots,a_r)\prod_{i=1}^{r-1}da_i}{\int_0^\infty\cdots\int_0^\infty \mathbb{E}\{K_1^{(1)}(t-\sum_{i=1}^{r}a_i)\}f_{1,\ldots,r}(a_1,\ldots,a_r)\prod_{i=1}^{r}da_i} \tag{19}$$

Hence, for the special case of the exponential state of a cell population at once follows the formula derived by MacDonald [32]

$$h_r(a_r)=\frac{\exp(-pa_r)\int_0^\infty\cdots\int_0^\infty \exp(-p\sum_{i=1}^{r-1}a_i)f_{1,\ldots,r}(a_1,\ldots,a_r)\prod_{i=1}^{r-1}da_i}{L_{1,\ldots,r}(p,\ldots,p)} ,$$

where the symbol $L_{1,\ldots,r}(p_1,\ldots,p_r)$ denotes the r-dimensional unilateral Laplace transform of the function f, i.e.

$$L_{1,\ldots,r}(p_1,\ldots,p_r)=\int_0^\infty\cdots\int_0^\infty \exp(-\sum_{i=1}^{r}p_ia_i)f_{1,\ldots,r}(a_1,\ldots,a_r)\prod_{i=1}^{r}da_i . \tag{20}$$

If the length of phase r does not depend on the durations of phases $1,2,\ldots,r-1$, (19) again easily reduces to formula (4) written only for the r-phase

$$h_r(a_r,t)=\mathbb{E}\{K_1^{(r)}(t-a_r)\}f_r(a_r)/\mathbb{E}\{K_2^{(r)}(t)\} .$$

Let us find the mean value of the distribution density $h_r(x_r)$

for the case of an exponentially growing population

$$\tilde{\mathbb{E}}\{X_r\}=\int_0^\infty x_r h_r(x_r)dx_r=\frac{\int_0^\infty\cdots\int_0^\infty x_r \exp(-px)f_{1,\ldots,r}(x_1,\ldots,x_r)\prod_{i=1}^{r} dx_i}{L_{1,\ldots,r}(p,\ldots,p)} . \quad (20')$$

By means of standard reasoning formula (20) at $p_1=p_2=\ldots=p_r=p$ is transformed to the form

$$L_{1,\ldots,r}(p,\ldots,p)=\int_0^\infty \exp(-px)f_{1+2+\ldots+r}(x)dx .$$

Transforming in a similar way the numerator of formula (20′), we obtain the following expression for the a posteriori mean value of the r-phase duration

$$\tilde{\mathbb{E}}\{X_r\}=\frac{\int_0^\infty\int_0^\infty x_r \exp(-px)f(x,x_r)dx_r dx}{\int_0^\infty \exp(-px)f_{1+2+\ldots+r}(x)dx},$$

or in a concise form

$$\tilde{\mathbb{E}}\{X_r\}=\mathbb{E}\{X_r e^{-pX}\}/\mathbb{E}\{e^{-pX}\} ,$$

where $X=\sum_{i=1}^{r} X_i$. Studying cell kinetics within the framework of a model of a stochastic age-dependent branching process, Jagers [26] introduced a more general characteristic $\tilde{\mathbb{E}}\{Y\}$ (flux expectation) for any random variable Y associated with the cycle of every cell

$$\tilde{\mathbb{E}}\{Y\}=\mathbb{E}\{Ye^{-pX}\}/\mathbb{E}\{e^{-pX}\},$$

by means of which it is possible to give a broader interpretation of the distribution density $h(y)$.

Associated with the concept of "flux expectation" is an important result obtained by Jagers [26] within the framework of an age-dependent branching process model. It may be formulated as follows: under the conditions of a steady exponential growth of a population the area B_1 under the first wave of FLM(t) curve is equal to

$$B_1=\tilde{\mathbb{E}}\{X_S(e^{pX_M}-1)\}/\tilde{\mathbb{E}}\{e^{pX_M}-1\}\simeq\tilde{\mathbb{E}}\{X_SX_M\}/\tilde{\mathbb{E}}\{X_M\},$$

where X_S and X_M are the durations of the S- and M- mitotic cycle phases, respectively.

If X_S and X_M and the length of the remaining portion of the cycle are mutually independent, then

$$B_1=\mathbb{E}\{X_Se^{-pX_S}\}/\mathbb{E}\{e^{-pX_S}\}\leq\mathbb{E}\{X_S\},$$

the equality attaining when and only when either $p=0$ or $\sigma_S^2=0$.

From the short survey of basic approaches to the mathematical modelling of cell kinetics presented here, it is evident that the exponential and steady-state (in strict sense) conditions allow to develop quite a natural description of FLM curve on the basis of "flux-expectations" approach. However, the solution of the general problem of analysing the FLM curve under arbitrary dynamic states of cell proliferation meets with a whole series of considerable difficulties.The main difficulty, naturally, is that one is dealing in the general case with a non-stationary function h(a,t) which rather complicates the choice of the appropriate approximation and completely cancels the advantage of a posteriori (flux-expectation method) modelling of FLM curves. In order to construct the FLM curve for more complicated population states, and in the first instance under conditions of transient cell kinetics, it is possible to use a different approach, consisting of a detailed reconstruction of the streams of labelled and unlabelled cells in each concrete situation. The dynamic phase structure of the cell cycle is then described directly by the function f(x) , reflecting the a priori distribution of the periods the cell spends in each group of transitory states.The mathematical model of the FLM curve cannot, in that case, be represented in a closed analytical form and requires additional information about other indices of cell kinetics and their change with time, obtained from an independent experiment. Of course, the use of additional functions, with their associated errors of measurement, must have an adverse effect on the accuracy with which the theoretical FLM curve is identified.

Another difficulty, arising when the simplest exponential state conditions do not hold, is caused by the indeterminate fate of descendent cells. Indeed, under real-life conditions, one does not usually know what fraction (even on average) of cells enters the mitotic cycle again after division, and what fraction goes into the resting state (G_0) , or begins to differentiate. This fact makes the correct evaluation of the number of times that labelled cells enter mitosis impossible. In such cases, the mathematical model describes only single entries into the mitotic phase and the results of the applied analysis relate only to the first wave of the experimental FLM curve.

Thus it should be expected that the modelling of a wider class of phenomena in cell kinetics is associated with the necessity of introducing some restrictions on the theoretical description of FLM curves and their applications. These restrictions also presume a type of problem which on the one hand is dictated by the performance of biological experiments, and which on the other hand is controlled by the corresponding methods of the mathematical analysis of the FLM curve. In this type of problem we can include the following:the study of the dependency of the FLM on the temporal parameters of the mitotic cycle phases of cells and the checking of existing graphical methods for the determination of these parameters; estimating the influence of various factors on the FLM shape,e.g., the state of the cell population at the time of introducing the label, and investigating the causes of certain anomalies in the experimental FLM; estimating the degree of correspondence between autoradiographical experimental data and contemporary ideas about the time sequence of events in the cell cycle; the study of possibilities of obtaining additional information through complex analysis of FLM curves and other experimental attributes of the kinetics of cell populations.

4.3. Mathematical Model Based on Transient Phenomena in Cell Kinetics

4.3.1. Introductory Formalism

Before embarking on the construction of the theoretical fraction of labelled mitoses curve under different states of cell proliferation processes, let us make some preliminary observations. We shall introduce the notion of integral streams (fluxes) of cells into some phase i of the mitotic cycle

$$r_i(t,t_0)=\int_{t_0}^{t} k_{1,i}(\tau)d\tau \qquad (21)$$

where t_0 is a fixed moment in time prior to the current moment t, and $k_{1,i}$ is the expected rate of entry of cells into phase i. Similarly we can define the stream of labelled cells, denoting it by $r^*(t,t_0)$. In the following, only the mathematical expectations of the stochastic processes will be considered. Let us turn again to the equation (1) and represent it in the form

$$n_i(t)=\int_0^t k_{1,i}(\tau)[1-F_i(t-\tau)]d\tau + v_i(t,0). \qquad (22)$$

Here

$$v_i(t,0)=\int_{-\infty}^{0} k_{1,i}(\tau)[1-F_i(t-\tau)]d\tau = \qquad (23)$$

$$n_i(0)\int_0^{\infty}\frac{[1-F_i(t+\tau)]}{[1-F_i(t)]} \quad w_i(\tau,0)d\tau,$$

$n_i(t)$ is the expected number of cells in phase i at moment t, $F_i(t) = \int_0^t f_i(u)du$ (where $f_i(u)$ is the probability density function of phase i length) and $w_i(\tau,0)$ is density of distribution of the age of cells which at time t=0 are in phase i. The fact that the second argument of the function v_i is zero means that the transient process is considered starting with t=0. In the case of an arbitrary starting point $t_0 = 0$, equation (22) takes the form

$$n_i(t)=\int_{t_0}^{t} k_{1,i}(\tau)[1-F_i(t-\tau)]d\tau+v_i(t,t_0) \quad (22')$$

where

$$v_i(t,t_0)=\int_{-\infty}^{t_0} k_{1,i}(\tau)[1-F_i(t-\tau)]d\tau = \qquad (23')$$

$$n_i(t_0)\int_0^{\infty} \frac{[1-F_i(t-t_0+\tau)]}{[1-F_i(\tau)]} w_i(\tau,t_0)\, d\tau.$$

Integrating (22) by parts and using the flux of cells of definition (21) we obtain an integral equation of the second kind for the function $r_i(t,0)$:

$$n_i(t)-v_i(t,0)=r_i(t,0)-\int_0^t r_i(\tau,0)f_i(t-\tau)\, d\tau \,. \qquad (24)$$

Similarly, equation (22') can be rewritten as

$$n_i(t)-v_i(t,t_0)=r_i(t,t_0)-\int_{t_0}^{t} r_i(\tau,t_0)f_i(t-\tau)d\tau \,. \qquad (25)$$

The solution of (24) can be expressed as follows:

$$r_i(t,0)=n_i(t)-v_i(t,0)+\int_0^t [n_i(\tau)-v_i(\tau,0)]\Psi_i(t-\tau)d\tau \,, \qquad (26)$$

where $\Psi_i(t-\tau)$ is the resolving kernel of the integral equation (24). If $f(u)$ can be approximated by the density of standard Γ-distribution with parameters α and β then on any finite interval $[0,t]$ the function $\Psi(t)$ can be obtained as the sum of the uniformly convergent series (see: Chapter III)

$$\Psi(t)= \sum_{k=1}^{\infty} \frac{\beta^{\alpha k}}{\Gamma(\alpha k)} e^{-\beta t} t^{\alpha k-1}, \quad \alpha\geq 1 \,, \qquad (27)$$

where

$$\alpha = \bar{\tau}_i^2/\sigma_i^2 \qquad \text{and} \qquad \beta = \bar{\tau}_i/\sigma_i^2 \,.$$

Thus, given the transient process $v_i(t,0)$, the stream $r_i(t,0)$

is uniquely determined for any finite value of $t \geq 0$.

Let us consider now an auxiliary problem - transform the flux $r_i(t,0)$ into the expected number of cells $n_{i+2}(t)$ which are in phase i+2. Assuming that, in phase i, the departure of cells to the G_0 state is impossible, the number of cells which enter phase i+1 in the interval 0 to t is equal to

$$r_{i+1}(t,0)=r_i(t,0)-n_i(t)+n_i(0). \tag{28}$$

From (28) and (24) we obtain

$$r_{i+1}(t,0)=\int_0^t r_i(\tau,0)f_i(t-\tau)d\tau + n_i(0)-v_i(t,0) \ . \tag{29}$$

Next we use the representation

$$r_i(t,t_0)=r_i(t,0)-r_i(t_0,0), \ 0<t_0<t \ ,$$

from which by taking (29) into account, it follows that in the absence of a G_0 state in phase i+1

$$r_{i+2}(t,t_0)=\int_0^t r_{i+1}(\tau,0)f_{i+1}(t-\tau)d\tau-\int_0^{t_0} r_{i+1}(\tau,0)f_{i+1}(t_0-\tau)d\tau - \tag{30}$$
$$v_{i+1}(t,0)+v_{i+1}(t_0,0) \ .$$

From the properties of the function $v_i(t,0)$, described earlier, it follows that for every positive ε_i , no matter how small, one can find $\delta_i(\varepsilon_i)>0$ such that $v_i(t,0)<\varepsilon_i$ when $t>\delta_i$. Therefore, by taking a suitable δ_{i+1}, which may be called the maximum duration of phase i+1 , and considering only the values $t>t_0>\delta_{i+1}$,we can ignore the last two terms in equation (30).

Then one can write

$$r_{i+2}(t,t_0)=\int_0^t r_{i+1}(\tau,0)f_{i+1}(t-\tau)d\tau-\int_0^{t_0} r_{i+1}(\tau,0)f_{i+1}(t_0-\tau)d\tau \ . \tag{31}$$

From (25), the number of cells in phase i+2 for $t>t_0$ is given by the formula

$$n_{i+2}(t)=r_{i+2}(t,t_0)-\int_{t_0}^t r_{i+2}(\tau,t_0)f_{i+2}(t-\tau)d\tau + v_{i+2}(t,t_0) \ , \tag{32}$$

and, assuming that $t>t_0+\delta_{i+2}>\delta_{i+1}+\delta_{i+2}$ we obtain finally

$$n_{i+2}(t)=r_{i+2}(t,t_0)-\int_{t_0}^{t} r_{i+2}(\tau,t_0)f_{i+2}(t-\tau)d\tau \ . \tag{33}$$

4.3.2. Construction of the FLM Curve in the General Case

Let phase i correspond to period S, phase $i+1$ to period G_2 and phase $i+2$ to period M in the mitotic cycle. We shall assume that the function $v(t,0)$ is either given or can be obtained from experimental data for only the S-phase. Then the following algorithm can be used to transform the number $n_S(t)$ of cells in phase S into the number $n_M(t)$ of cells in phase M:

1. The function $n_S(t)$ must be given on the entire period of observation $[0,T]$;
2. The integral equation (24) must be solved by some means (e.g., through formulae (26) and (27)) to give the function $r_S(t,0)$ for $t \in [0,T]$;
3. On the same interval $[0,T]$ the stream of cells from phase S into phase G_2, $r_{G_2}(t,0) = r_S(t,0)-n_S(t)+$ $+n_S(0)$ must be computed (see (28));
4. By fixing a moment of time $t_0>\delta_{G_2}$, the function $r_M(t,t_0)$ is evaluated for $t\geq t_0$ by using formula (31):

$$r_M(t,t_0)=\int_0^t r_{G_2}(\tau,0)f_{G_2}(t-\tau)d\tau-\int_0^{t_0} r_{G_2}(\tau,0)f_{G_2}(t_0-\tau)d\tau \ ;$$

5. For $t \in (\delta_{G_2}+\delta_M, T]$ the number of cells in phase M is computed according to formula (33)

$$n_M(t)=r_M(t,t_0)-\int_{t_0}^{t} r_M(\tau,t_0)f_M(t-\tau)d\tau.$$

Thus, in order to describe the functions $n_M(t)$ on the interval $(\delta_{G_2}+\delta_M, T]$ one has to know the behaviour of $n_S(t)$ on the larger interval $[0,T]$. Furthermore, that domain for $n_S(t)$ may also turn out to be insufficient because, in order to determine $v_S(t,0)$ for $t>0$ (see the following) one needs to know the history of the process prior to the moment $t=0$, at least as far back as the interval $(-\delta_S, 0)$. Therefore, only by having this extra information about the function $n_S(t)$ is it possible to determine the relative number of mitoses with time, using the above algorithm.

The situation simplifies significantly when one considers the processes of induced cell proliferation; then the initial numbers of cells in phases S, G_2 and M may be ignored and, by setting $v_S(t,0) \equiv v_{G_2}(t,0) \equiv v_M(t,0) \equiv 0$, the algorithm for transforming $n_S(t)$ into $n_M(t)$ can be rewritten as follows:

$$
\begin{aligned}
&(1) \quad n_S = n_S(t) \quad \text{for} \quad t \in [0,T];\\
&(2) \quad r_S(t,0) = n_S(t) + \int_0^t n_S(\tau)\Psi_S(t-\tau)d\tau;\\
&(3) \quad r_{G_2}(t,0) = r_S(t,0) - n_S(t);\\
&(4) \quad r_M(t,0) = \int_0^t r_{G_2}(\tau,0) f_{G_2}(t-\tau)d\tau;\\
&(5) \quad n_M(t) = r_M(t,0) - \int_0^t r_M(\tau,0) f_M(t-\tau)d\tau.
\end{aligned}
\tag{34}
$$

It is now easy to obtain similar expressions for the expected number of labelled mitoses $n_M^*(t)$ and the fraction of labelled mitoses $FLM(t) = n_M^*(t)/n_M(t)$. Let $t=0$ correspond to the moment of introduction of the impulse label and let labelling be incorporated only in those cells which were in phase S at $t=0$. Then during a certain period after $t=0$ (i.e., until the second DNA sythesis of previously labelled cells begins) only unlabelled cells will enter into phase S. If one exclusively follows that population of cells which were impulse labelled at $t=0$ through the mitotic cycle, then the procedure of introducing the label is formally equivalent to the block $G_1 \rightarrow S$. Therefore, by setting $r_S(t,0)=0$ in (29) we obtain for $t \geq 0$

$$
r^*_{G_2}(t,0) = n_S(0) - v_S(t,0), \tag{35}
$$

and from this point onwards exactly the same transformations of the stream of labelled cells $r^*_G(t)$ as in steps 3)-5) of algorithm (34) can be applied. As a result, the number of labelled mitoses is given by

$$n_M^*(t)=r_M^*(t,0)-\int_0^t r_M^*(\tau,0)f_M(t-\tau)d\tau \qquad (36)$$

where

$$r_M^*(t,0)=\int_0^t r_{G_2}^*(\tau,0)f_{G_2}(t-\tau)d\tau,$$

and the function $r_{G_2}^*(t,0)$ is defined by (35). After the sequence of operations necessary for the determination of the functions $n_M^*(t)$ and $n_M(t)$, given $n_S(t)$ and $v_S(t,0)$, the construction of the FLM curve is completed by computing the function

$$FLM(t)=n_M^*(t)/n_M(t).$$

Under experimental conditions, instead of the absolute number of cells in one or another phase of the cell cycle, one usually operates with the phase index $I_i(t)=n_i(t)/n_\Sigma(t)$, where $n_\Sigma(t)$ is the total number of cells in the system being studied. Of practical significance are the indices of phases S (labelling index) and M (mitotic index) which are equal, by definition, to

$$I_S(t)=n_S(t)/n_\Sigma(t) \qquad \text{and} \qquad I_M(t)=n_M(t)/n_\Sigma(t).$$

When studying the processes of induced or stimulated cell proliferation it is convenient to introduce the notion of modified indices (see: Section 3.4)

$$\tilde{I}_S(t)=n_S(t)/n_\Sigma(0) \qquad \text{and} \qquad \tilde{I}_M(t)=n_M(t)/n_\Sigma(0).$$

In exactly the same way, in the analysis of experimental data, it is necessary to replace the function $r_i(t,t_0)$ by the indicator

$$q_i(t,t_0)=r_i(t,t_0)/n_\Sigma(t_0) \quad . \qquad (37)$$

In cases where $n_\Sigma(t)$ can be considered constant, e.g., under strict stationary conditions, or diurnal variation in the processes of cell proliferation, replacing the absolute quantities $n_i(t)$ by the relative indicators $I_i(t)$ does not change anything in the computation of $FLM(t)=I_M^*(t)/I_M(t)$. If, however, the reproduction and death of cells influence significantly the total number of cells in the system then this needs to be taken into account. The simplest approximation method which allows for these factors in kinetic processes of stimulated proliferation employs the formula

$$\tilde{I}_i(t)=I_i(t)\exp\{\frac{1}{\bar{\tau}_M}\int_0^t [I_M(x)-\bar{\tau}_M\varkappa(x)]dx\},$$

where $\varkappa(t)$ is the intensity of the random (i.e., independent of the phase in the mitotic cycle) death of cells.

The theoretical methods for studying FLM curves can be simplified significantly if the analysis is applied not to the curve $FLM(t)$ but to the index of labelled mitoses $I_M^*(t) = n_M^*(t)/n_\Sigma(t)$. It is possible to obtain the index $I_M^*(t)$, i.e. the proportion of labelled mitoses among all cells in a given system,by multiplying the experimental $FLM(t)$ and $I_M(t)$. This, however, increases the error in the data being analysed. Nevertheless, the use of $I_M^*(t)$ in the analysis of experimental material can frequently be justified [17,31] especially in cases where a rough estimate of the mean mitotic duration $\bar{\tau}_M$ is required. Indeed if we let $n_\Sigma(t)$=constant, and there is no prolonged delay of cells in the S- and G_2-phases (e.g., on account of exit into G_0) and if we can assume further that $I_M^*(t)$ shows only single entries of labelled cells in mitosis, then the area under the experimental curve I_M^* must be equal to $I_S(0)\bar{\tau}_M$:

$$\int_0^\infty I_M^*(t)dt=\frac{1}{n_\Sigma}\int_0^\infty k_{1,M}^*(\tau)d\tau\int_\tau^\infty [1-F_M(t-\tau)]dt =$$

$$\frac{1}{n_\Sigma}\int_{-\infty}^t k_{1,M}^*(t-x)dx\int_0^\infty [1-F_M(x)]dx = \tag{38}$$

$$\frac{\bar{\tau}_M}{n_\Sigma}\int_0^\infty k_{1,M}^*(t)dt=I_S(0)\bar{\tau}_M.$$

Under stricter steady-state conditions, i.e., when $n_S(t)=$ constant and $n_M(t)=$constant a graphical method for the evaluation of $\bar{\tau}_S$ also follows from expression (38) [14]

$$\int_0^\infty FLM(t)dt = \bar{\tau}_S \quad .$$

The latter relation may also be obtained from the Jagers theorem considered in Section 4.2.

In the above arguments it was assumed that the function $v_S(t,t_0)$ (or the ratio $v_S(t,t_0)/n_\Sigma(t)$) is given a priori. Let us proceed now to methods of obtaining it from the experimental data on $I_S(t)$. First of all, from the way the functions $r_S(t,t_0)$ and $v_S(t,t_0)$ are defined (see (21) and (23')) we can write

$$v_S(t,t_0)=r_S(t_0,-\infty)[1-F_S(t-t_0)]-\int_{-\infty}^{t_0} r_S(\tau,-\infty)f_S(t-\tau)d\tau, \; t\geq t_0. \qquad (39)$$

Returning to (23'), it is easy to see that $v_S(t_0,t_0)=n_S(t_0)$. However in order to determine $v_S(t,t_0)$ for $t>t_0$ it is necessary, as can be seen from (39), to determine the function $r_S(t,-\infty)$ for $t \in (-\infty,t_0)$ which in turn, requires knowledge of the function $n_S(t)$ on the same interval of time. This interval can be reduced by use of truncated distributions. One is thus forced to abandon the construction of $v_S(t,t_0)$ and $r^*_{G_2}(t,t_0)$ in the general case and to consider only some special states of cell kinetics, namely those which are most frequently encountered in practice.

4.3.3. Induced DNA Synthesis

We start by noting that the moment when the proliferative stimulus begins to act corresponds to $t=0$ and the ^{3}H-thymidine is introduced later at some moment $t_0>0$. If at $t=0$ the number of cells in phase S of the mitotic cycle is negligibly small, then in describing the induced transfer of cells into DNA synthesis it can be assumed that $n_S(0)=0$ and, therefore, $v_S(t,0)\equiv 0$. The initial conditions at time t_0 and, therefore, the

transient process $v_S(t,t_0)$, are determined by the cells which have entered phase S of the mitotic cycle by that moment after the initial action of the proliferative stimulus. Substituting $r_S(0,-\infty)=0$
into (39) we obtain

$$v_S(t,t_0)=r_S(t_0,0)[1-F_S(t-t_0)]-\int_0^{t_0} r_S(\tau,0)f_S(t-\tau)d\tau, \quad t\geq t_0, \qquad (40)$$

where the function $r_S(t,0)$ on the interval $[0,t_0]$ is the solution of the integral equation (see: (24))

$$n_S(t,0)=r_S(t,0)-\int_0^t r_S(\tau,0)f_S(t-\tau)d\tau. \qquad (41)$$

In the early stages of induced cell proliferation it is almost always possible to choose t_0 such that, together with the increase in the labelling index, there will be sufficiently low mitotic activity to justify the assumption $n_\Sigma(t_0)=n_\Sigma(0)$. Then dividing both sides of (40) and (41) by $n_\Sigma(0)$ and bearing in mind equations (26) and (37), we obtain for the function $\tilde{v}_S(t,t_0)=v_S(t,t_0)/n_\Sigma(0)$ the following expression:

$$\tilde{v}_S(t,t_0)=q_S(t_0,0)[1-F_S(t-t_0)]-\int_0^{t_0} q_S(\tau,0)f_S(t-\tau)d\tau , \qquad (42)$$

where

$$q_S(t,0)=\tilde{I}_S(t) +\int_0^t \tilde{I}_S(\tau)\Psi_S(t-\tau)d\tau . \qquad (43)$$

Having obtained the function $\tilde{v}_S(t,t_0)$, we can compute the modified mitotic labelling index $\tilde{I}^*_M(t)$, by substituting the corresponding relative indicators into formula (36).

In real systems with induced cell proliferation (see, e.g., [12,13]) it is quite common to be able to set the value of t_0 according to the condition $q_S(t,0) \cong I_S(t)$ for $t \in [0,t_0]$. Then, using equations (35), (37) and (42), we can write the following simple expression for the relative stream of labelled cells into phase G_2:

$$q^*_{G_2}(t,t_0)=r^*_{G_2}(t,t_0)/n_\Sigma(0)=I_S(t_0)F_S(t-t_0)+\int_0^{t_0} I_S(\tau)f_S(t-\tau)d\tau \ , \quad (44)$$

which, in turn, can be used for constructing $\tilde{I}^*_M$ in accordance with algorithm (34). In order to obtain the index $\tilde{I}_M(t)$ and hence $FLM(t)=\tilde{I}^*_M(t)/\tilde{I}_M(t)$, it is necessary to determine $q_S(t,0)$ on the entire observation interval with the aid of (43). The above considered algorithm was applied to analysis of FLM(t) of PHA-stimulated normal and irradiated lymphocytes [51]. Another application to analysis of liver regeneration will be described in Chapter V.

4.3.4. A Cell Population Synchronized in the S-Period

If at the moment $t=0$ only cells of zero-th age are in the S-phase and their number is equal to $n_S(0)$, then according to considerations of Section 3.6 the function $v_S(t,0)$ is described by the expression

$$v_S(t,0)=n_S(0)[1-F_S(t)].$$

Suppose that the labelled precursor is also introduced at the moment $t=0$ and subsequently only the synchronized portion of the cell population is involved in the proliferation processes. In that case it is clear that the flux of labelled cells into the G_2-phase is $r^*_{G_2}(t,0)=n_S(0)F_S(t)$. If after the zero moment in the S-period new cells continue to enter it, the total flux of cells into the G_2-phase will be equal to

$$r_{G_2}(t,0)=\int_0^t r_S(\tau,0)f_S(t-\tau)d\tau+n_S(0)F_S(t)$$

or

$$q_{G_2}(t,0)=\int_0^t q_S(\tau,0)f_S(t-\tau)d\tau \ + \ I_S(0)F_S(t).$$

In order to find $q_S(t,0)$ note that the function $v_S(t,0)$ in this case satisfies the integral equation

$$v_S(t,0) + \int_0^t v_S(\tau,0)\Psi_S(t-\tau)d\tau=n_S(0),$$

from which, in view of (26), follows

$$q_S(t,0)=\tilde{I}_S(t) - \tilde{I}_S(0) + \int_0^t \tilde{I}_S(\tau,0)\Psi_S(t-\tau)d\tau .$$

Consideration of a more general case of synhronization of cells at any other finite age $a_0>0$ of the S-period would present no particular problems (see Chapter III).

4.3.5. The Strict-Sense Steady State and the Diurnal Rhythm of Cell Proliferation

Description of the integral flux of labelled cells $r^*_{G_2}(t,0)$ for the steady state follows at once from formula (35) and formula (78) in Chapter III

$$r^*_{G_2}(t,0)=\frac{n_S}{\bar{\tau}_S}\int_0^t [1-F_S(x)]dx.$$

For an influx related to the total mean number of cells in a population we have the expression

$$q^*_{G_2}(t,0)=\frac{I_S}{\bar{\tau}_S}\int_0^t [1-F_S(x)]dx ,$$

which, using the initial distribution of cell age , may be written as

$$q^*_{G_2}(t,0)=I_S\int_0^t w_S(a,0)da . \qquad (45)$$

The meaning of expression (45) becomes clear when it is considered that under steady-state conditions the distribution function for the residual cell lifetime in the S-period, i.e. $1-v_S(t,0)/n_S(0)$, coincides with the age distribution function $\int_0^t w_S(a,0)da$.

The procedure for constructing the function $v(t,t_0)$ for diurnal rhythm of cell proliferation was outlined in Section 3.6. Thus, in this case too the flux of labelled cells into the G_2-phase may be calculated according to formula (35) and then all transformations may be performed which are described in the foregoing algorithm of constructing a theoretical labelled mitoses curve.

In the preceding subsection a method was presented for simulating repetitive FLM waves in exponentially growing cell populations based on the assumption that after mitosis both daughter cells must invariably enter at once a new cycle of division. The use of such an assumption (or defining the mean fraction of daughter cells entering the next mitotic cycle) makes it possible to describe the entry of labelled cells once again into mitosis also under the conditions of unsteady-state cell kinetics by means of the method of cell flux transformations outlined in the foregoing. However, in many cases, specifically in studying systems with induced cell proliferation, either the assumption of the closed cell population structure may fail to be met or the mean fraction of daughter cells involved in the next mitotic cycle is a composite time function which cannot be satisfactorily allowed for in FLM analysis. Therefore, concrete applications of FLM analysis in systems of unsteady-state cell kinetics are generally confined to search for estimates of temporal parameters (mean value and variance of cycle phase duration) of the S,G_2- and M-periods of the mitotic cycle by comparing the first experimental and theoretical FLM(t) waves. Temporal parameter estimates can be obtained, for example, by means of the least squares method and non-linear programming algorithms.

4.4. Investigation of Labelled Mitoses Curve Behaviour under Unsteady-State Cell Kinetics Conditions

The present section is concerned with the results of studying the effect of transient process dynamics on the form of the first FLM wave. This investigation was carried out by conducting model experiments on a computer in which arbitrary variants of transient processes were formed. Thus, the principal aim of the investigation consisted in reproducing diverse states of cell kinetics and the peculiarities of FLM which may occur in such states. Since in simulating FLM on a computer, the values of numerical parameters of cycle phase duration distributions are defined by the investigator, parameter estimates obtained by the graphical method were verified at the same time. Comparison of estimates obtained graphically and by means of a mathematical model was also undertaken in references [7,46] but it was then confined to the state of a population with a stationary cell age distribution in the cycle. The method for constructing theoretical FLM with fixed temporal parameters of mitotic cycle phases, presented in the preceding subsection, makes possible a detailed comparison under the conditions of a considerable variety of states of cell proliferation kinetics.

In order to reveal peculiarities of FLM behaviour in model experiments on a computer a certain reference state of cell kinetics should be defined so that variations in the structure of FLM coincident with different deviations from that state could be investigated.

In our study the state selected was a strict-sense steady state of cell proliferation kinetics for which the relative flux of cells into the S-phase is

$$q_S(t) = \frac{I_S}{\bar{\tau}_S} t .$$

Thus, the steady state conditions are provided by defining $q_S(t)$ as a linear time function and selecting the moment of label introduction t_0 sufficiently remote from the zero moment. In our experiments the moment was usually selected from the condition $t_0 \geq 2\bar{\tau}_S$; the time the stationary state was reached in the mitotic phase being determined by the mean duration of the G_2-phase.

<u>Simulation experiment 1.</u> The purpose of the experiment was to find out the manner in which FLM(t) varies with cell proliferation state before and after label introduction at the moment t_0. The following values of temporal parameters of the S-, G_2- and M-phases of the mitotic cycle were used (hrs):
$\bar{\tau}_S=10$, $\sigma_S=2$, $\bar{\tau}_{G_2}=2$, $\sigma_{G_2}=0{,}5$, $\bar{\tau}_M=1$, $\sigma_M=0{,}3$.

In the calculations which follow it was assumed that cycle phase durations obey the Γ-distribution.

Represented in Fig.6 are different $q_S(t)$ functions underlying construction of FLM(t) with the fixed pulse-labelling moment $t_0=20$ hr (the OCFKN straight line defines the strict-sense steady state).

At the selected temporal parameter values the stationary state in the M-phase with all $q_S(t)$ definition variants would set in within 20 hr. With the first four variants (Fig.6,b), which simulated changes in the kinetics of cell entry into the S-phase prior to the moment of labelling, the dynamics of $q_S(t)$ after the moment t_0 was identical for all the variants and corresponded to the steady state (the FKN portion of the OCFKN control straight line). It is seen from Fig.6,b that the FLM(t) curve responds to changes in the dynamic prehistory of the S-phase, such changes affecting only the descending branch of the first FLM wave. The following regular pattern can be traced: the curvilinear portions of the $q_S(t)$ plot lying below the OCF portion of the control straight line appear in the flatter descending FLM branch and conversely , the curvilinear $q_S(t)$ portions above OCF are associated with a more rapid decay of FLM(t). At the same time to more distinct departures of the $q_S(t)$ function from the OCF control state there correspond more pronounced deviations in the behaviour of the descending portions of the FLM(t) curve. As may be seen from Fig.6,c, the behaviour of FLM(t) also varies with cell proliferation state after the moment $t_0=20$ hr. Graphical estimation [40] of the mean length of the S-period from the 50 (or 37) percent FLM level with deviations from the strict-sense steady state may lead to blunders. The graphical method (by the 50 percent level) yields good estimates of true parameter values for the stationary state: $\bar{\tau}$ =10 hr, $\bar{\tau}$ =2,5 hr.

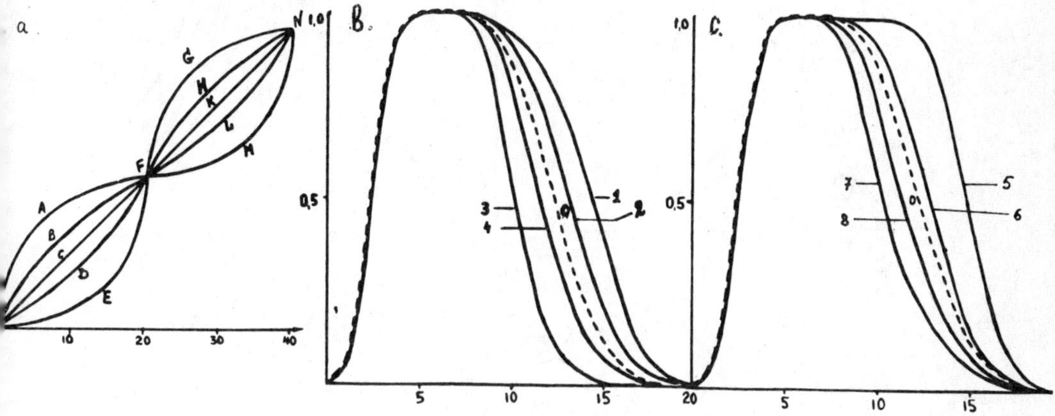

Figure 6. Experiment 1. FLM(t)-q_S(t) relationship before (b) and after (c) label introduction. "a"-q_S(t) values (%) "b" and "c"- FLM(t) values corresponding to q_S(t) definition variants: 0 - OCFKN, 1 - OEFKN, 2 - ODFKN, 3 - OAFKN, 4 - OBFKN, 5 - OCFMN, 6 - OCFLN, 7 - OCFGN, 8 - OCFHN.

<u>Simulation experiment 2.</u> All the conditions of experiment 1 were preserved but the mean duration of the G_2-phase was extended considerably. The set of temporal parameters was as follows: $\bar{\tau}_S$=10 hr, σ_S=2 hr, $\bar{\tau}_{G_2}$=7 hr, σ_{G_2}=3 hr, $\bar{\tau}_M$=0,9 hr, σ_M=0,3 hr. The steady state set in at ~30 hr. The results of the experiment are presented in Fig.7. It may be seen that with the longer G_2-phase variations in FLM(t) with those in q_S(t) prior to the labelling instant t_0=20 hr are more pronounced and occur throughout the whole length of the first FLM wave. Variations in q_S(t) after the moment t_0 also give rise to deviations of FLM(t) more marked than in experiment 1 but affecting only the descending branch of the curve. Otherwise the variations are of the same character as in experiment 1. The graphical method for

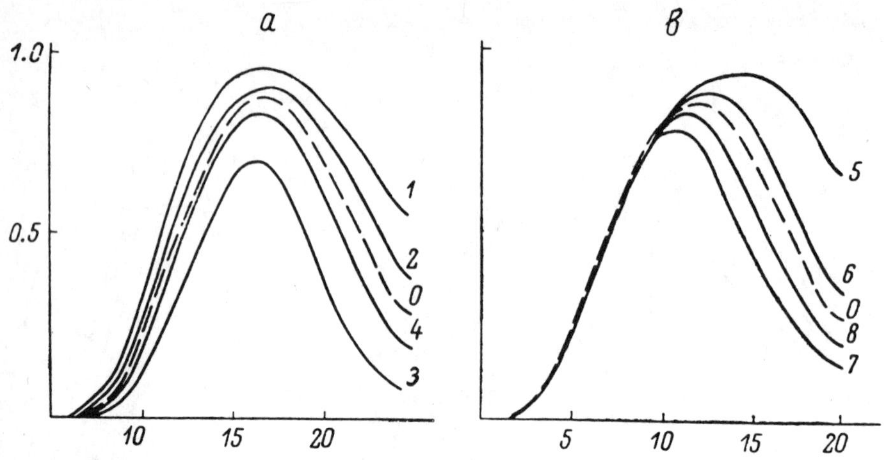

Figure 7. Experiment 2: same as in experiment 1 (Fig.6) at $\bar{\tau}_{G_2}$=7 hr and σ_{G_2}=3 hr. Designations are the same as in Fig.6 b and c.

determining the mean durations of the S- and G_2-phases (by the 50 percent FLM level) yields accurate estimates $\bar{\tau}_S$=10 hr, $\bar{\tau}_{G_2}$=7 hr for the steady state, for other variants such estimates may, indeed, differ markedly from true values. With a prolonged G_2-phase FLM may fail to reach the 100 percent level.

<u>Simulation experiment 3.</u> This experiment was concerned with investigating the role of plateau in the $q_S(t)$ curve which, as shown in the next chapter, is quite a real feature of cell proliferation processes. Represented in Fig.8 is the function $q_S(t)$ with a plateau extension commensurable with the S-phase

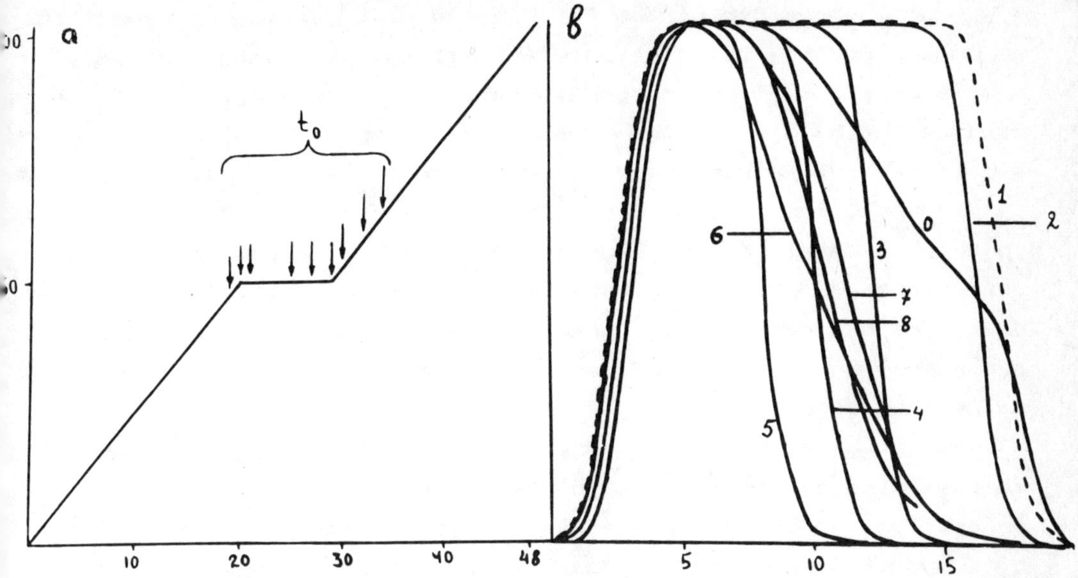

Figure 8. Experiment 3. Relationship between FLM(t) form and the moment of labelling in the presence of a plateau in the $q_S(t)$ curve. a- $q_S(t)$. The arrows indicate different moments of label introduction t_0 ; b - FLM(t) curves at different t_0 : 0-t_0=19 hr, 1- t_0=20 hr, 2-t_0=21 hr, 3-t_0=25 hr, 4-t_0=27 hr, 5-t_0= 29 hr, 6-t_0= 30 hr, 7-t_0=32 hr, 8-t_0=34 hr.

mean duration and equal to 9 hr. The temporal parameters of mitotic cycle phases are the same as in experiment 1. It will be seen from Fig.8 that the introduction of label at different time moments t_0 gives rise to a variety of FLM forms. Apparently, at t_0=19 hr we have one of the variants investigated in the second part of experiment 1. Similar results were obtained by means of stochastic simulation [50].

A new pattern manifests itself with t_0 ranging between 20 and 29 hr. When the label is introduced at the moment t_0=20 hr corresponding to the beginning of the plateau in $q_S(t)$ the widest FLM(t) curve results. Further increase in t_0 within the plateau area leads to a concurrent shift of the descending FLM(t) branch along the temporal axis towards lower values. At the same time replacement of t_0 by $t_0+\Delta$ within the range specified above brings about a decrease in the 50 percent width of FLM by the quantity Δ. For example, the 50 percent width of FLM(t) at t_0=29 hr is precisely 9 hr less than the FLM(t) width at

t_0=20 hr . The width of the FLM plateau also decreases, but the ascending portion is invariable for all the variants (Fig.8). A similar pattern is noted with variations in the duration of cell contacts with ^{3}H-thimidine [5]. In our case the introduction of labels at different moments in the presence of a plateau in the $q_S(t)$ curve corresponds formally to variations in time interval in which a labelled precursor is incorporated in the DNA of cells entering the S-phase. Variants of FLM(t) calculation for t_0>29 hr demonstrate a considerably more complex character of FLM(t) dependence on the state of cell proliferation at the moment of pulse labelling.

Simulation experiment 4. One of the possible after-effects of ^{3}H-thymidine on the progress of cells through the mitotic cycle (radiation and metabolic effects) may be variations in the values of temporal parameters of some cycle phases for labelled and unlabelled cells. For simulation studies of the problem calculation of the $\tilde{I}^*_M(t)$ function (i.e. FLM(t) numerator) was carried out with altered parameters of the G_2-phase, the set of parameters of all cycle phases required for computing $\tilde{I}_M(t)$ (FLM(t) denominator) remained unchanged. Strict-sense steady state alone was investigated; in some of the variants the mean duration of the G_2-phase was increased or diminished at a fixed coefficient of variation $\sigma_{G_2}/\bar{\tau}_{G_2}$, while some others were specifically concerned with elucidating the role of variance. It has been found that variations in $\bar{\tau}_{G_2}$ lead to FLM(t) shifts along the temporal axis: with $\bar{\tau}_{G_2}$ decreasing or increasing by the quantity Δ the whole FLM(t) curve shifts by the same quantity to the left or the right, respectively. A two-fold variation of σ_{G_2} produced no appreciable effect on the FLM(t) behaviour. Determination of $\bar{\tau}_{G_2}$ by the 50 percent FLM yields quite satisfactory estimates of the true value of that parameter for labelled cells. It may be inferred that the information contained in FLM on the progress of cells through the G_2-period in the steady state refers primarily to labelled cells.

Supplementary notes

Testing the graphical method for estimating $\bar{\tau}_S$ and $\bar{\tau}_{G_2}$ in a large series of additional experiments have shown that satisfactory estimates from the 50 percent FLM level can be usually obtained only when cell population states are characterized by stationary cell-age distribution in a cycle.

Under real conditions of a biological experiment anomalies of the behaviour of the first FLM wave are not infrequent. For instance, fitting into such anomalies is a "crevasse" located roughly in the middle of the first FLM and noted mainly in experimental systems with non-stationary age distributions. The crevasse's depth varies considerably in size. The causes responsible for that phenomenon remain obscure, yet certain authors are inclined to attach to it a profound biological meaning. Thus, it has been suggested [18] that in some systems, approaching the middle of the S-phase, the cells interrupt DNA synthesis (for about 1 hr) which leads to the formation in some sense of two S-periods separated by a short time interval. Of course, if we include this assumption in a mathematical FLM model, a two-peak FLM can be constructed. However, such a crevasse may also be regarded as an anomaly of FLM behaviour accounted for by more natural causes. In a set of simulation experiments it was attempted to reproduce a two-peak structure of the first FLM(t) wave by selecting such $q_S(t)$ functions which should cause cell accumulations at the beginning and end of the S-period by the moment of label introduction. In none of the situations a crevasse in the FLM(t) curve could be reproduced. On the other hand, the following explanation may be offered for the phenomenon in question.

It should be emphasized once again that, as a rule, no crevasse manifests itself in the peak of the first FLM wave in experiments performed on cell systems in a strict-sense steady or exponential state. On the other hand, to obtain every point in an experimental FLM use is always made of different elements of the population under study (different animals, different slides with a cell monolayer, etc.) which, under the conditions of unsteady-state kinetics, may differ greatly in the dynamic state of the S-phase at the moment of labelling. Thus, the variability of experimental data due to random factors is superimposed on by the variability in respect to the initial state. The results of the simulation experiments presented above show that the contribution of that factor may be decisive in the origin of different anomalies of FLM behaviour and, consequently, adequate interpretation of an experimental FLM is possible only when one has comprehensive data on the dynamic state of proliferation processes in the cell system under study.

The effect of transient processes on the form of FLM in irradiated cell populations was studied by computer simulation in [53].

4.5. Labelled Mitoses Curve under the Conditions of the Diurnal Rhythm of Cell Proliferation Processes

It was pointed out in Chapter III that diurnal variations in the labelling and mitotic indices may be accounted for by fluctuations of the rate of cell entry into the DNA synthesis or/and changes in the durations of mitotic cycle phases dependent on the time of day. As regards the former explanation it may be assumed, for example, that periodic blocking of cells occurs at some point (dichophase) of the G_1-phase and the cells remain outside the mitotic cycle while the blocking factor is in effect. In that case the time the cells are in the block (outside the cycle) is not included in the duration of the G_1-phase. After termination of the block the cells continue their progress through the cycle from the point where they were stopped. If the time the cells stay in the block is included in the duration of the G_1-phase, we should, naturally, consider a non-stationary G_1-phase. Thus, as justly pointed out by Macdonald [34] in discussing the work by Yakovlev et al. [52], the two models are not mutually exclusive. The alternative formulated above refers only to such isolated transitive populations (e.g., corresponding S-, G_2- and M-phases of the mitotic cycle) whose stationary character we wish to verify and is but a convenient way of using the advantages inherent in the concept of a transitive cell population. An investigation based on such methodological principles and described in what follows corroborates the view expressed in the literature [37] that involved in the diurnal rhythm of proliferative processes are both factors: partial synchronization of cells and periodic fluctuations in the lengths of mitotic cycle phases.

By way of proving the existence of a periodic trend in the temporal parameters of mitotic cycle phases in cell systems with a marked diurnal rhythm of proliferative processes reference is usually made to the fact that labelled mitoses curves obtained with labels introduced at different times of day may differ to a considerable extent [3,4,25,38,42]. All data of that kind were obtained by graphically estimating parameters of mitotic cycle phases using the Quastler and Sherman method [40], no account being taken of the effect which the dynamics of transient processes of cell kinetics exerts on the form of FLM. The results of simulation experiments (see the preceding subsection) suggest that, with a distinct diurnal rhythm in cell proliferation processes, FLM corresponding to different times of day the labels were introduced at may differ according to a regular pattern in their characteristics even when the temporal parameters of the S-, G_2- and M-phases are not subject to diurnal variations. The first attempt to explore the problem by means of mathematical modelling was made by Klein and Valleron [30] who modified for this purpose the method for formalizing cell kinetics proposed by Takahashi [47-49]. The modification was as follows. Representing the mitotic cycle as a set of unilaterally transitive states in accordance with the principle of Takahashi's method, the authors formulated the sinusoidal law of transition probability changes for the substates which form the G_1-phase, considering that transition probabilities for the states constituting the other phases (S,G_2 and M) independent of the current time. Thus, included in modified Takahashi's model are periodic variations in the mean duration only of the G_1-phase of the mitotic cycle. Klein and Valleron used the Takahashi's model thus modified for analyzing experimental data of Izquierdo and Gibbs [25] on diurnal fluctuations of the values of the labelling index $I_S(t)$ and the mitotic index $I_M(t)$, as well as FLM in the hamster cheek pouch epithelium for different moments of ^{3}H-thymidine introduction (8 and 20 hr). The experimental data are presented in Fig.9. The most conspicuous result of the analysis performed by Klein and Valleron was the construction of two different in form (mainly descending branches) FLM(t) for t_0=8 hr and t_0=20 hr although, as stated above, the temporal parameters of the S-, G_2- and M-phases were fixed. The two FLM(t) curves showed a satisfactory agreement with real observed FLM.

For the purpose of comparing the approaches Yakovlev et al. [52] also turned to the same experimental data, approximating the $I_S(t)$ curve by the method of least squares with the following function

$$I_S(t)=0.041 + 0.025 \sin \frac{\pi\ (t-3.909)}{12} \quad .$$

Optimization of estimates of the mean values and variances of the S-, G_2 and M-phase durations was carried out by means of a simultaneous minimization of the sum of squares of deviations of the $I_M(t)$ and FLM(t) theoretical curves from their experimental counterparts on the basis of the Nelder-Mead method [22]. First of all the set of parameters obtained by Klein and Valleron in fitting together $I_S(t)$ and $I_M(t)$ curves was tested, i.e. $\bar{\tau}_S$=10 hr, σ_S=2 hr, $\bar{\tau}_{G_2}$=1,75 hr, σ_{G_2}=0,56 hr, $\bar{\tau}_M$=1,3 hr, σ_M= 0.58 hr. The kinetic curves constructed at these values (represented by dashed lines in Fig.9), though differing somewhat from those obtained by Klein and Valleron, demonstrate the same pattern: labelling at different times of day give rise to different forms of FLM which, apparently, is due to time-of-day variations in cell age distribution in the S-phase of the mitotic cycle. Optimization search for temporal parameters by I(t) and FLM(t) curves at t_0=8 hr led to the following estimates: $\bar{\tau}_S$=8.73 hr, σ_S=2.04 hr, $\bar{\tau}_{G_2}$=1.76 hr, σ_{G_2}=0.6 hr, $\bar{\tau}_M$=1.31 hr and σ_M=0.19 hr. This set of parameters improves considerably the agreement between the $I_M(t)$ and FLM(t)

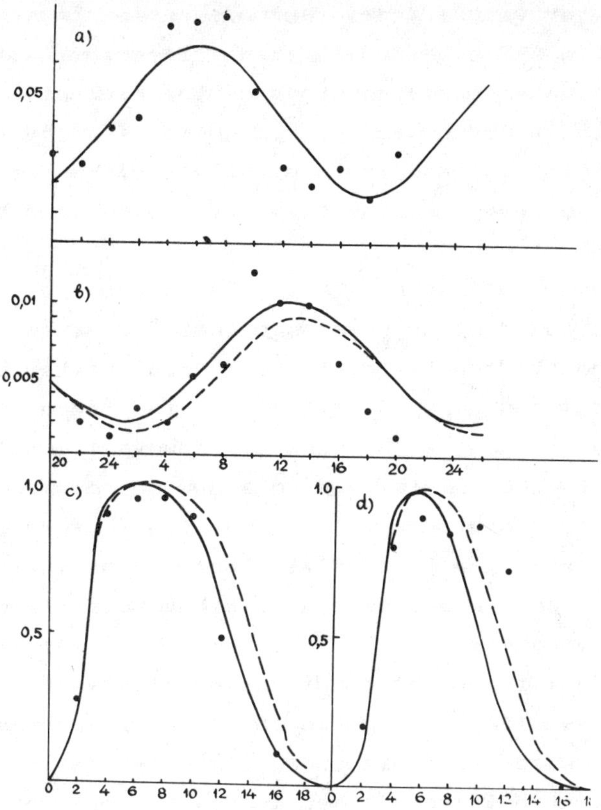

Figure 9. Differences in FLM in the hamster cheek pouch epithelium with label introduction at 8 hr (c) and 20 hr (d) under the conditions of diurnal rhythm of cell proliferation. On the abscissa: a,b-the times of day(hr);c,d-time after labelling(hr); on the ordinate: a-the labelling index,b-the mitotic index;c-,d-fraction of labelled mitoses. The dots represent the experimental data [25];the dashed lines denote the theoretical curves at the parameter values according to [30]; the solid lines are approximation of $I_S(t)$ by the least-squares technique (a) and the corresponding $I_M(t)$ and FLM(t) curves constructed at parameter values obtained for the case of label introduction at 8 hr (b,c,d).

theoretical curves at t_0=8 hr and the experimental data as compared with the set of parameters recommended by Klein and Valleron [30] . However, agreement with FLM at t_0=20 hr with these parameter values indeed becomes worse. Thus, in that case differences in FLM obtained for two different moments of label introduction under the conditions of the diurnal rhythm of cell proliferation cannot be fully accounted for by the effect of transient processes, and the possibility of a certain trend of cycle phase durations in the time of day appears to be quite distinct.

It stands to reason that the procedure for parameter optimization could be so developed that the search for estimates would be conducted on the basis of a joint criterion of closeness of theoretical and experimental FLM simultaneously for t_0=8 hr and t_0=20 hr. Such a procedure, however, would add further uncertainty to the problem, while a new set of parameters obtained by it would not necessarily be universal for any other t_0 values. The existence of diurnal fluctuations in the duration of the S-, G_2- and M-phases has also been confirmed by Clausen et al.[5].

There is still another complication involved in solving the problem of revealing a trend in the temporal parameters of mitotic cycle phases associated with the time of day. Introduction of ^{3}H-thymidine at different times of day is followed by dissimilar dynamics of the $I_S(t)$ and $I_M(t)$ curves [4]. Moller [36] has shown that irrespective of all other external conditions an injection of ^{3}H-thymidine induces a higher mitotic activity of cells. This feature is not taken into account in the experimental data of Jzquierdo and Gibbs [25] and, consequently, was not allowed for in the calculations by Yakovlev et al. [52] although its effect on the results of mathematical analysis of FLM under the conditions of diurnal rhythm of cell proliferation may appear quite appreciable.

It is, indeed, much easier to establish the existence of a periodic trend in the temporal parameters of cycle phases than to quantitatively estimate it. Aimed at solving that latter problem are more intricate models of non-stationary cell systems whose theory and applications are dealt with by a number of authors [16,21,23,24,28,29,33,34]. The methods for constructing, investigating and applying such models are discussed in detail in Macdonald's work [34].

REFERENCES

1. Barrett, J.C. A mathematical model of the mitotic cycle and its application to the interpretation of percentage labelled mitoses data , J.Nat.Cancer Inst., 37, 443-450,1966.
2. Barrett, J.C. Optimized parameters of the mitotic cycle, Cell Tiss.Kinet., 3, 349-353,1970.
3. Burns, E.R. and Scheving, L.E. Circadian influence on the wave form of the frequency of labeled mitoses in mouse corneal epithelium, Cell Tiss.Kinet., 8,61-66, 1975.
4. Burns, E.R., Scheving, L.E., Fawcett, D.F., Gibbs, W.W. and Galatzan, R.E. Circadian influence on the frequency of labeled mitoses method in the sratified squamous epithelium of the mouse esophagus and tongue, Anat. Res., 184, 265-274, 1976.
5. Dondua, A.K. and Dondua, G.K. On mitotic cycle analysis, In: Studies in cell cycles and metabolism of nucleic acids in differentiation, Nauka, Moscow, Leningrad, 5-36, 1964 (In Russian).
6. Clausen, O.P.F, Thorud, E., Bjerknes, R. and Elgjo, K. Circadian rhythms in mouse epidermal basal cell proliferation, Cell Tiss.Kinet., 12, 319-337, 1979.
7. Denecamp, J. The cellular proliferation kinetics of animal tumors, Cancer Res., 30, 393-406, 1970.
8. Dombernowsky, P., Bichel, P. and Hartmann, N.R. Cytokinetic analysis of the JB-1 ascites tumor at different stages of growth, Cell Tiss.Kinet., 6, 347-357, 1973.
9. Dombernowsky, P. and Hartmann, N.R. Analysis of variations in the cell population kinetics with tumor age in the L1210
10. Dörmer, P.,Brinkmann, W.,Born, R. and Steel, G.G. Rate and time of DNA synthesis of individual Chinese hamster cells, Cell Tiss. Kinet., 8, 399-412, 1975.
11. Eisen, M. Mathematical models in cell biology and cancer chemotherapy, Springer-Verlag, Berlin, Heidelberg, New York, 1979.
12. Ellem, K.A.O. and Mironescu, S. The mechanism of regulation of fibroblastic cell replication.I. Properties of the system, J.Cell.Physiol., 79, 389-406, 1972.
13. Fabrikant, J.I. The kinetics of cellular proliferation in regenerating liver, J.Cell Biol., 36, 551-565, 1968.
14. Gerecke,D. An improved method for the evaluation of DNA synthesis time from the graph of labelled mitoses, Exper. Cell Res., 62, 487-489, 1970.
15. Gilbert, C.W. The labelled mitoses curve and the estimation of the parameters of the cell cycle, Cell Tiss.Kinet.,5, 53-65, 1972.
16. Guiguet, M.,Klein, B. and Valleron, A.J. Diurnal variation and the analysis of percent labelled mitoses curves, In:Biomathematics and Cell Kinetics,Elsevier/North-Holland Biomed. Press, Amsterdam, 191-1978.
17. Guschin, V.A. A mathematical model for the cell system kinetics of the stratified squamous epithelium of the hamster cheek pouch, Cytology, 13, 1426-1432, 1971 (In Russian).
18. Hamilton, A.I. Cell population kinetics: a modified interpretation of the graph of labeled mitoses, Science, 164, 952-954,

1969.

19. Hartmann, N.R. and Pedersen, T. Analysis of the kinetics of granulosa cell population in the mouse ovary, Cell Tiss. Kinet., 3, 1-11, 1970.

20. Hartmann, N.R., Gilbert, C.M., Jansson, B., Macdonald, P.D.M., Steel, G.G. and Valleron, A.J. A comparison of computer methods for the analysis of fraction labelled mitoses curves, Cell Tiss.Kinet., 8, 119-124, 1975.

21. Hartmann, N.R. and Møller, U. A compartment theory in the cell kinetics including considerations on circadian variations, In: Biomathematics and Cell Kinetics, Elsevier / North-Holland Biomed.Press, Amsterdam, 223-251, 1978.

22. Himmelblau, D.M. Applied Nonlinear Programming, McGraw-Hill Book Company, 1972.

23. Hopper, J.L. and Brockwell, P.J. Analysis of data from cell populations with circadian rhythm, In: Biomathematics and Cell Kinetics, Elsevier/North-Holland Biomed. Press, Amsterdam, 211-221, 1978.

24. Hopper, J.L. and Brockwell, P.J. A stochastic model for cell populations with circadian rhythms, Cell Tiss. Kinet., 11, 205-225, 1978.

25. Izquierdo, J.N. and Gibbs, S.J. Turnover of cell-renewing populations udergoing circadian rhythms in cell proliferation, Cell Tiss.Kinet., 7, 99-111, 1974.

26. Jagers, P. Branching Processes with Biological Applications, Wiley, London, 1975.

27. Karle, H., Ernst, P. and Killman, S. Changing cytokinetic patterns of human leukaemic lymphoblasts during the course of the disease, studied in vivo, Brit.J.Haematol., 24, 231-244, 1973.

28. Klein, B. and Guiguet, M. Relative importance of the phases of the cell cycle for explaining diurnal rhythms in cell proliferation in the tissues with a long G_1 duration, In: Biomathematics and Cell Kinetics, Elsevier / North-Holland Biomed.Press, Amsterdam, 199-210, 1978.

29. Klein, B. and Macdonald, P.D.M. The multitype continuous-time Markov branching process in a periodic environment, Adv.Appl. Prob., 12, 81-93, 1980.

30. Klein, B. and Valleron, A.J. Mathematical modelling of cell cycle and chronobiology: preliminary results, Biomedicine, 23, 214-217, 1975.

31. Liosner, L.D. and Markelova, J.V. The mitotic cycle of regenerating liver hepatocytes, Bull.Exper.Biol.Med., 71, 99-103, 1971 (In Russian).

32. Macdonald, P.D.M. Statistical inference from the fraction labelled mitoses curve, Biometrika, 57, 489-503, 1970.

33. Macdonald, P.D.M. Age distributions in the general cell kinetic model, In: Biomathematics and Cell Kinetics, Elsevier/North-Holland Biomed.Press, Amsterdam, 3-20, 1978.

34. Macdonald, P.D.M. Measuring circadian rhythms in cell populations, In: The Mathematical Theory of the Dynamics of Biological Populations II, Academic Press, London, 1981.

35. Malinin, A.M. and Yakovlev, A.Yu. The labeled mitoses curve in different states of cell proliferation kinetics, Cytology, 18, 1270-1277, 1976 (In Russian).

36. Møller, U. Interaction of external agents with the circadian mitotic rhythm in the epithelium of the hamster cheek pouch,

J.Interdiscipl. Cycle Res., 9,105-114, 1978.
37.Møller,U. and Larsen, J.K. The circadian variations in the epithelial growth of the hamster cheek pouch: quantitative analysis of DNA distributions, Cell Tiss. Kinet., 11, 405-413,1978.
38.Møller,U., Larsen, J.K. and Faber,M. The influence of injected tritiated thymidine on the mitotic circadian rhythm in the epithelium of the hamster cheek pouch, Cell Tiss.Kinet.,7, 231-239,1974.
39.Pedersen,T. and Hartmann, N.R. The kinetics of granulosa cells in developing follicules in the mouse ovary,Cell Tiss.Kinet., 4, 171-184, 1971.
40.Quastler,H and Sherman, F.H.Cell population kinetics in the intestinal epithelium of the mouse, Exper. Cell Res., 17, 429-438, 1959.
41.Scheufens, E.E. and Hartmann, N.R. Use of gamma distributed transit times and the Laplace transform method in theoretical cell kinetics, J.Theoret.Biol., 37, 531-543,1972.
42.Sigdestad, C.P. and Lesher, S. Circadian rhythm in the cell cycle time of the mouse intestinal epithelium,J.Interdiscipl. Cycle Res., 3, 39-46, 1972.
43.Steel, G.G. The cell cycle in tumours: an examination of data gained by the technique of labelled mitoses, Cell Tiss. Kinet.,5,87-100, 1972.
43.Steel, G.G. The measurement of the intermitotic period, In: The Cell Cycle in Development and Differentiation,Cambridge,13 - 29,1973.
45.Steel, G.G. Growth Kinetics of Tumours, Oxford, 1977.
46.Steel, G.G. and Hanes, S. The technique of labelled mitoses: analysis by automatic curve-fitting,Cell Tiss.Kinet., 4, 93-105, 1971.
47.Takahashi, M. Theoretical basis for cell cycle analysis. I. Labeled mitosis wave method, J.Theoret.Biol., 13, 202-212, 1966.
48.Takahashi,M. Theoretical basis for cell cycle analysis. II. Further studies on labeled mitosis wave method, J.Theoret.Biol.,18, 195-209, 1968.
49.Takahashi, M., Hogg, G.D. and Mendelsohn, M.L. The automatic analysis of PLM-curves, Cell Tiss.Kinet., 4,505-518, 1971.
50.Toivonen,H. and Rytömaa, T. Monte Carlo simulation of malignant growth, J.Theoret.Biol.,78, 257-267, 1978.
51.Yakovlev, A.Yu. Kinetics of proliferative processes induced by phytohemagglutinin in irradiated lymphocytes,Radiobiology,23, 449-453, 1983 (In Russian).
52.Yakovlev, A.Yu., Lepekhin, A.F. and Malinin,A.M. The labeled mitoses curve in different states of cell proliferation kinetics. Y. The influence of diurnal rhythm of cell proliferation on the shape of the labeled mitoses curve, Cytology, 20, 630-635, 1978 (In Russian).
53.Yakovlev, A.Yu. and Zorin, A.V. Computer Simulation in Cell Radiobiology, Springer-Verlag, Berlin, Heidelberg, New York, 1988.

V. APPLICATIONS OF KINETIC ANALYSIS. RAT LIVER REGENERATION.

5.1. Introduction

From the exposition in Chapter III it follows that a broader scope of kinetic analysis of induced cell proliferation may eventuate from investigating the peculiarities of behaviour of $q_S(t)$ and $P_S(t)$ indices which make it possible to separately assess the processes of initial and recurrent transition of cells to DNA synthesis after the onset of the effect of a proliferative stimulus. For calculating $q_S(t)$ and $P_S(t)$ it is necessary to have experimental data on the dynamics of such indices as $I^c_S(t)$, continuously labelled cells index, $I_S(t)$, pulse labelled cells index, and $I_M(t)$, mitotic index, as well as estimates of the temporal parameters $\bar{\tau}_S$, σ_S and $\bar{\tau}_M$. In certain cases the parameters $\bar{\tau}_S$ and σ_S may be evaluated without the labelled mitoses curve (FLM) , an instance being given in Chapter III. The method for estimating the parameters of the S-, G_2- and M-phases of the mitotic cycle described in Chapter IV is based on optimization of the parameters of theoretical FLM(t) whose construction, in turn, includes as an indispensable step calculation of the q-index for the S-phase. In this way estimating the temporal parameters and constructing the q_S- index are combined in a single computation procedure.

The set of experimental data on the dynamics of the principal kinetic indices: $I^c_S(t)$, $I_S(t)$, $I_M(t)$ and FLM(t) required for kinetic analysis of systems with induced or stimulated proliferation (SISP) may be found in Fabricant's works [11-16] dealing with regeneration of the liver of MacCollum and August strain male rats. Fabrikant's experimental findings will be drawn upon in this chapter to illustrate the application of the methods described above for analysis of induced cell proliferation and subsequent discussion of peculiarities inherent in proliferative reaction of parenchymal hepatic cells to partial hepatectomy (PHE)-excision of 2/3 of hepatic tissue.

5.2. Kinetic Analysis of Induced Hepatocyte Proliferation in Regenerating Rat Liver

Analyzing his findings, Fabrikant drew the following conclusions:

(1) Over 85 percent of quiescent (with respect to mitosis) hepatocytes in the intact liver are capable of DNA synthesis and division in response to PHE.

(2) The percentage index of labelled hepatocytes begins to grow in regenerating liver at least 12 hr after PHE. After a rapid increase in DNA synthesis at 22 hr in a relatively large number of hepatocytes, the rate of cell entry into the S-phase remains practically unchanged within the next 28 hr to be followed by a gradual decline in the rate of cell transition to DNA synthesis.

(3) The duration values for the S-phase and M-phase of the mitotic cycle of hepatocytes, measured by standard graphic methods, using the labelled mitoses curve are 8 and 1 hr, respectively.

(4) An enhanced mitotic activity is observed at least 8 hr after a rise in DNA synthesis. Most of hepatocytes pass through mitosis once; however few cells undergo two divisions and more. Some centrolobular polyploid hepatocytes do no synthesize DNA during regeneration at all.

(5) The active proliferation of hepatocytes does not entail the death of any fraction of the cell population.

Let us consider first Fabrikant's experimental data on the MacCollum rats. A substantial shortcoming of those data is the absence of FLM curve, therefore estimation of the temporal parameters of the S-phase of the mitotic cycle of hepatocytes may be performed only by fitting the initial portion of the curve $q_S(t)$, i.e. until 24 hr, to the experimental values of $I_S^c(t)$ and $I_S(t)$. To obtain the required estimates it is necessary to define distribution of the S-phase of the mitotic cycle. Yakovlev et al. [50] used the Γ-distribution in analyzing the kinetics of induced DNA synthesis in the regenerating liver of the MacCollum rats according to Fabrikant's data. They obtained the following estimates of the temporal parameters of the S-phase: $\bar{\tau}_S$=10 hr and σ_S=2.5 hr.

In the present chapter we shall first present the results of study [50] and then pass to a detailed consideration of the radioautographic data on regeneration of the liver in the August strain rats. In so doing, we shall also make use of the Γ-distribution for describing the durations of all mitotic cycle phases. The choice of that parametric family of distributions is dictated above all by considerations of convenience since a number of studies [9,19, 49,51] have demonstrated that the results of kinetic analysis of cell proliferation are highly stable to the choice of approximation among different unimodal continuous distributions. The validity of the choice is also supported by time-lapse cinemicrography of cells in vitro [22,25,26,32].

Shown in Table 2 are the numerical results obtained in [50] from investigating the sensibility of indices $q_S(t)$ and $\Lambda(t) = n_\Sigma(t)/n_\Sigma(0)$ to value $\bar{\tau}_M$ variations. By means of the comparison of the function $\Lambda(t)$ which describes the increase in the total number of hepatocytes, following PHE, with the index $q_S(t)$ it is ipossible to estimate the lower limit of mean mitotic duration as 0.75 hr. Thus, Table 2 shows that the actual value of $\bar{\tau}_M$ lies within 0.75-2 hr. The $q_S(t)$ index does not respond much to insignificant variations in $\bar{\tau}_M$ during the first 40 hr of the liver regeneration, as far as real values of $\bar{\tau}_M$ go.

Fig.10 shows the curves of $\tilde{I}_S(t)$ and $q_S(t)$ plotted for two extreme values of $\bar{\tau}_M$=0.75 and 2 hr. Until 26 hr after PHE the curves show a mass cell transition to DNA synthesis; beginning from 26 hr after PHE, the rate of cell entry into the S-phase drops considerably, i.e. nearly to zero for $\bar{\tau}_M$=2 hr , and after 32 hr it starts to grow again. During 40 hr nearly 92% of the initial number of hepatocytes enter the S-phase for $\bar{\tau}_M$=0.75 hr and 80% for $\bar{\tau}_M$=2 hr. Cell transition to DNA synthesis after 25 hr occurs

Table 2.

Hours after PHE	$\bar{\tau}_M$ = 0.25 hr.		$\bar{\tau}_M$=0.70 hr.		$\bar{\tau}_M$=0.75 hr.		$\bar{\tau}_M$=0.8 hr.		$\bar{\tau}_M$= I hr.		$\bar{\tau}_M$= 2 hr.		$\bar{\tau}_M=\infty$
	q_s	Λ	q_s	Λ	q_s	Λ	q_s	Λ	q_s	Λ	q_s	Λ	q_s
I6	8.0	I.00	8.0	I.00	8.0	I.00	8.0	I.00	8.0	I.00	8.0	I.00	8.0
I8	I4.2	I.00	I4.2	I.00	I4.2	I.00	I4.2	I.00	I4.2	I.00	I4.2	I.00	I4.2
20	2I.9	I.00	2I.9	I.00	2I.9	I.00	2I.9	I.00	2I.9	I.00	2I.9	I.00	2I.9
22	34.I	I.02	33.8	I.0I	33.7	I.00	33.7	I.00	33.7	I.00	33.7	I.00	33.6
24	48.8	I.07	46.8	I.03	46.7	I.02	46.7	I.02	46.5	I.02	46.I	I.0I	45.7
26	64.2	I.I8	58.6	I.06	58.4	I.06	58.3	I.05	57.8	I.04	56.8	I.02	55.8
28	74.0	I.36	63.5	I.I2	63.I	I.II	62.8	I.I0	6I.9	I.08	60.I	I.04	58.5
30	84.8	I.67	67.4	I.20	66.8	I.I9	66.4	I.I7	65.0	I.I4	62.4	I.07	59.9
32	I00.0	2.24	72.5	I.33	7I.7	I.3I	7I.0	I.29	69.I	I.22	65.5	I.II	62.2
34	I27.5	3.I8	83.4	I.5I	82.3	I.47	8I.3	I.43	78.5	I.33	73.6	I.I5	69.2
36	I48.7	4.37	88.5	I.69	87.I	I.63	85.9	I.58	82.5	I.45	76.5	I.20	7I.4
38	I66.6	5.62	92.3	I.85	90.7	I.78	89.3	I.7I	85.4	I.54	78.6	I.24	72.9
40	I77.2	6.84	94.I	I.99	92.4	I.90	90.9	I.82	86.8	I.62	79.6	I.27	73.5

Dependence of functions $q_s(t)$ (in %) and $\Lambda(t)$ on mean mitotic duration $\bar{\tau}_M$.

Figure 10.

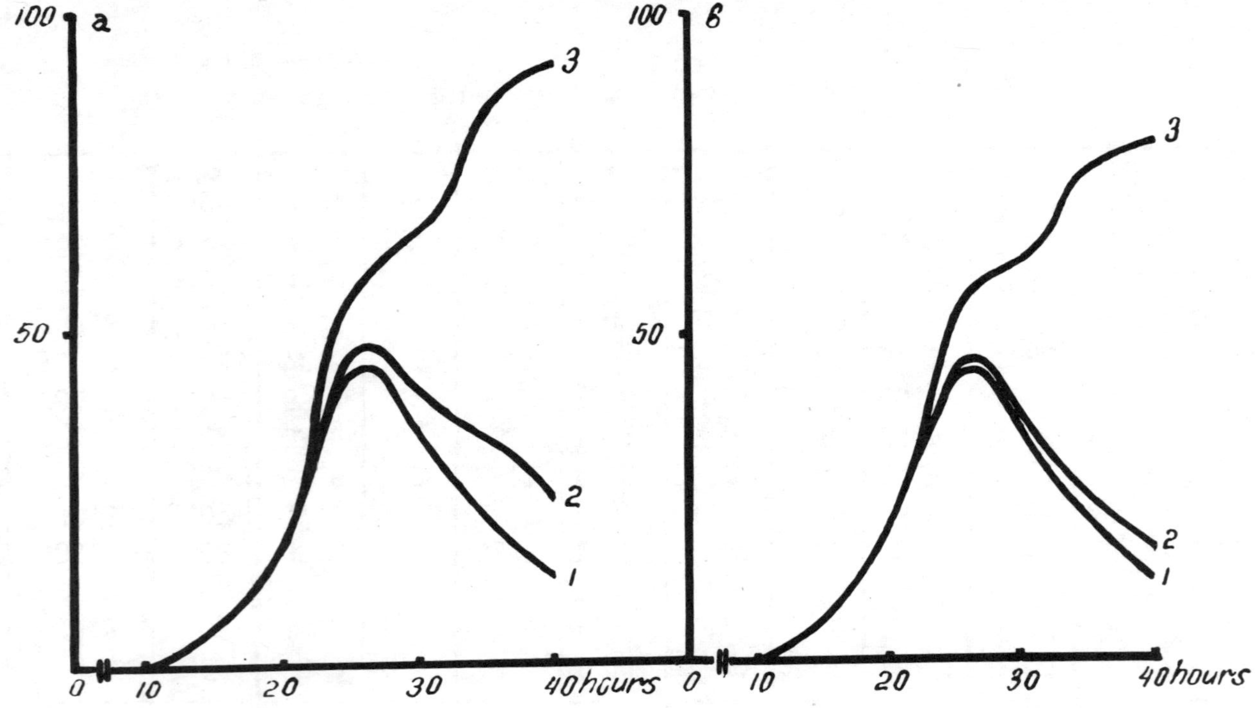

Temporal patterns of main indices of parenchymal cell proliferation in the regenerating liver of the Mac Collum rats. 1 - $I_S(t)$, 2 - $\tilde{I}_S(t)$, 3 - $q_S(t)$: (a) $\overline{\tau}_M$ =0.75 hr.; (b) $\overline{\tau}_M$ =2 hr.

when the mitotic activity is already high and the function $\tilde{I}_S(t)$ exceeds $I_S(t)$, beginning from that moment. The substantial decrease in the transition rate in the interval from 26 to 32 hr after PHE suggests that considerable fractions of cells enter the prereplicative period virtually synchronously (parasynchronously), namely 70% for $\bar{\tau}_M$=0.75 hr or 65% for $\bar{\tau}_M$=2 hr.

The hypothesis of a synchronous entry of the initial cell fraction into the prereplicative period may be tested on the basis of the formula

$$q_S(t) = C_0 F_{PS}(t) \tag{1}$$

where C_0 is the mean portion of cells in the synchronous fraction, $F_{PS}(t)$ is the function of distribution of the whole prereplicative period length covering the transformation period and the true G_1-phase of the mitotic cycle. Formula (1) appears to remain valid only for the time interval during which the function $q_S(t)$ represents transition to DNA synthesis only of the initial fraction of cells, i.e. 32 hr after PHE. Throughout that period of time a high accuracy of constructing $q_S(t)$ is ensured [49]. The $q_S(t)/C_0$ curves are quite satisfactorily approximated by the functions of Γ-distribution (Fig.11). The parameters of approximating functions chosen as estimates of the temporal parameters of the whole prereplicative period are τ_{PS}=22 hr and σ_{PS}=4.5 hr, respectively.

Investigation into the process of the first entry and re-entry of cells into DNA synthesis, i.e. the indices $P_S(t)$ and $H_S(t)$, also reveals a sudden entry of a considerable initial fraction of cells into the prereplicative period. On the basis of the index $I_S^C(t)$ presented in Fabrikant's work [15] the indices $P_S(t)$ and $H_S(t)$ for two terms of the liver regeneration, i.e. 25 and 37 hr, may be obtained (Table 3). Table 3 shows that it is at $\bar{\tau}_M$=2 hr only that the size of the synchronized fraction indicated by P_S(37 hr) equalling 61.5% is close to that determined from the plateau position on the $q_S(t)$ curve, i.e. ≅ 65%. The latter value

Figure 11.

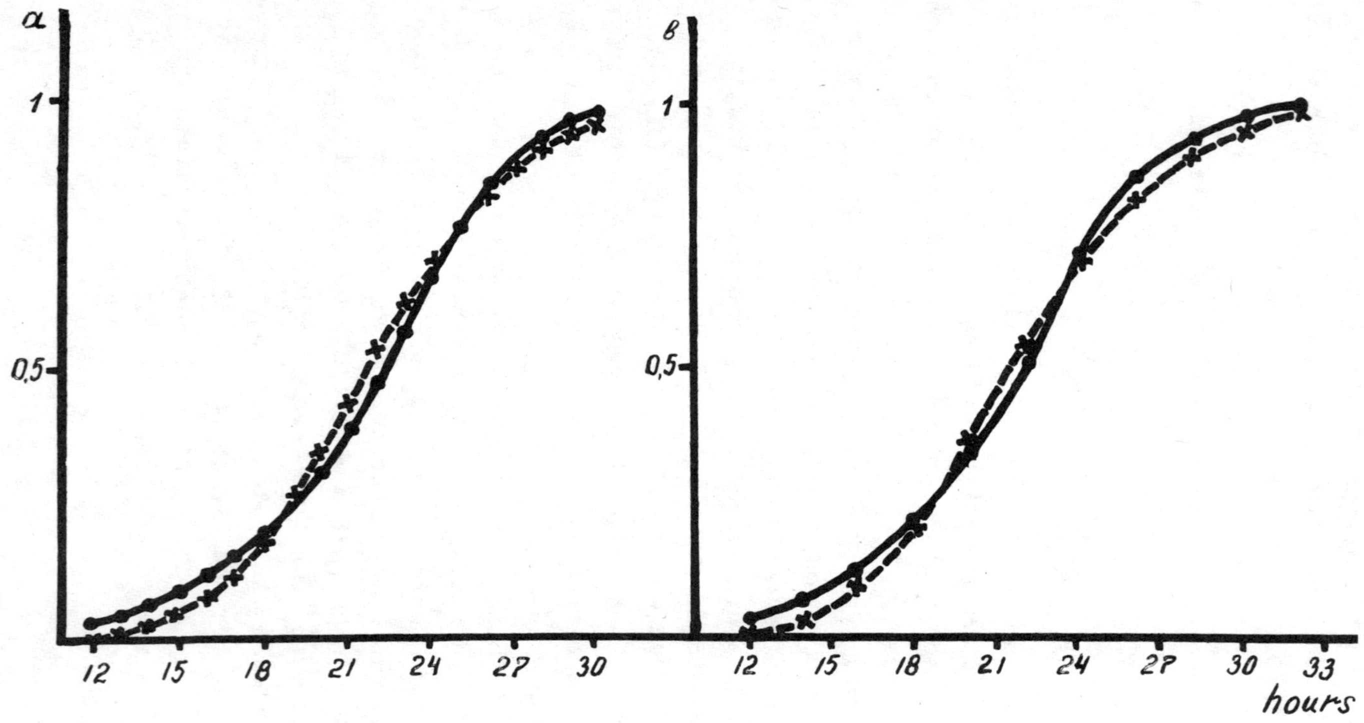

Test of hypothesis on parasynhronous entry of initial fraction of hepatocytes into the prereplicative period. Solid line shows curve $q_S(t)/C_o$, dashed line-approximation by Γ-distribution. (a) Computations for $\overline{\tau}_M$=0.75 hr.; (b) computations for $\overline{\tau}_M$=2 hr.

Table 3.

Initial and repeated entry of cells into the S-phase in the regenerating rat liver (MacCollum strain)

Hours after PHE	$I_S^c(t)$	$\bar{\tau}_M$=0.75hr		$\bar{\tau}_M$= 1hr		$\bar{\tau}_M$= 2hr	
		$P_S(t)$	$H_S(t)$	$P_S(t)$	$H_S(t)$	$P_S(t)$	$H_S(t)$
25	47%	42%	11%	46%	7%	47%	4%
37	68,5%	46%	42%	54%	30%	61,5%	15%

seems to be somewhat high, since at 25–30 hr after PHE the re-entry of hepatocytes (about 4–5%), contributions to the increase of the $q_S(t)$ function, into the S-phase should be expected. Hence, assuming the mean mitotic duration to be close to 2 hr, the size of the initial fraction of cells C_0 is estimated to be about 60% . All these considerations apparently proceed from the assumption that there are no particular features in the process of liver regeneration, such as, for example, the death of a fraction of the cell population or polyploidizing division, which may account for an uncontrolled drop in the value of $P_S(t)$ (or increase in $q_S(t)$). Therefore, the possibility of the value of $\bar{\tau}_M$ being actually somewhat less than 2 hr cannot be ruled out. Yager et al. [46] also point out that at least 56% of hepatocytes are involved in the first wave of DNA synthesis after PHE, neither do they rule out the possibility of later transition to proliferation by another fraction of the cell population.

The data presented in Table 3 also enable preliminary comparison of the temporal organizations of the first and second mitotic cycles of hepatocytes. It is evident from the Table 3 that with $\bar{\tau}_M$=2 hr at 37 hr after PHE about 15% of cells re-enter the S-phase. The passage of the 15% of cells through the whole prereplicative period takes longer than 18 hr (see Table 2). It means that unless the prereplicative period in the second mitotic cycle is shortened, 15% of the initial number of cells can not re-enter the S-phase at 37 hr. A similar analysis of the liver regeneration at 25 hr after PHE leads to the same conclusion. Consideration of the data in Table 3 for lower values of $\bar{\tau}_M$ yields still more convincing results. Therefore, there is every reason to believe that on the average the first entry of the cell into the S-phase takes longer than re-entry. This provides support for the hypothesis of the existence of a transformation period in the structure of the first mitotic cycle of hepatocytes. Indeed, it seems reasonable to assume that some time of the DNA synthesis induction is spent on the cells getting prepared to proliferate. For this reason the prereplicative period should not be identified with the G_1-phase of the mitotic cycle. As simple estimates show the minimal duration of the transformation period exceeds 6 hr. There is no transformation period in the structure of the second mitotic cycle of the initial fraction cells.

Thus, the kinetic analysis of the initial period (until 40 hr after PHE) of liver regeneration in rats of the MacCollum strain reveals the following main characteristics of hepatocyte proliferation:

(1) parasynchronous entry into the mitotic cycle of the initial fraction of hepatocytes (equal to about 60% of the entire hepatocyte population remaining in the liver after the operation);

(2) early entry of some hepatocytes into the second mitotic cycle;

(3) the existence of a special transformation period in the first mitotic cycle of hepatocytes following PHE.

Thus, the foregoing data indicate that only 60% of the initial number of hepatocytes rapidly enter the mitotic cycle in response to PHE, while 30% of cells become involved in the proliferative response after a significant delay. These results may be interpreted in two different ways.

(1) Original population of hepatocytes includes a fraction (60% of cells) with a high degree of competence for proliferation. Highly differentiated hepatocytes comprising the other part of the cell population take much more time to switch over their metabolism than the "rapid" hepatocytes from the 60%-fraction. In

other words,a certain portion (30%) of parenchymal cell population passes through an additional period of transformation (or dedifferentiation) with a minimal duration of over 25 hr.

(2). On the contrary, a population of hepatocytes is quite homogeneous in respect to readiness for DNA synthesis and mitosis, however, under the conditions of the experiment, i.e. with a 2/3 hepatectomy, the proliferative stimulus acts, on the average, only on 60% of cells. In the meantime the rest of the parenchymal cell population takes up an increased functional load.

Evidence in support of the latter interpretation of the results of kinetic analysis will be presented in what follows.

Non-uniformity of the cell system response to the proliferative stimulus complicates considerably the problem of interpreting the results of a biochemical investigation into liver regeneration in the mammals. Indeed, it may always be expected that some of the molecular events and metabolic shifts recorded in the course of liver regeneration are accounted for by the fraction of a cell population which ,at the time of observation, is not in the state of active proliferation.

The fact that not all the elements of a hepatocyte population are simultaneously involved in liver regeneration is further confirmed by data reported in the literature. For instance, Rixon and Whitfield [33] by means of the "suicide" technique presented a qualitative proof of the existence of two fractions of a parenchymal cell population which begin synthesizing DNA at different times following PHE. The authors ascertained two peaks both in the index of labelled cells curve and in that of mitotic index. Large doses of ^{3}H-thymidine introduced at the time of the first peak of DNA sythesis suppressed the first wave of the mitotic activity, without practically affecting the second. This observation has led Rixon and Whitfield to the conclusion that two proliferation waves are formed by subpopulations of hepatocytes characterized by different rates of development of proliferative reactions. Cells entering the mitotic cycle in separate (synchronous) fractions were also noted in other SISP, for instance, during regeneration of the rat lens epithelium [23], during stimulation of DNA synthesis in confluent cultures of diploid fibroblasts of the human lung [10] and Chinese hamster fibroblast - like cells [48]. The kinetics of phytohemagglutinin-induced proliferation of human peripheric blood lymphocytes is also characterized by that feature [47].

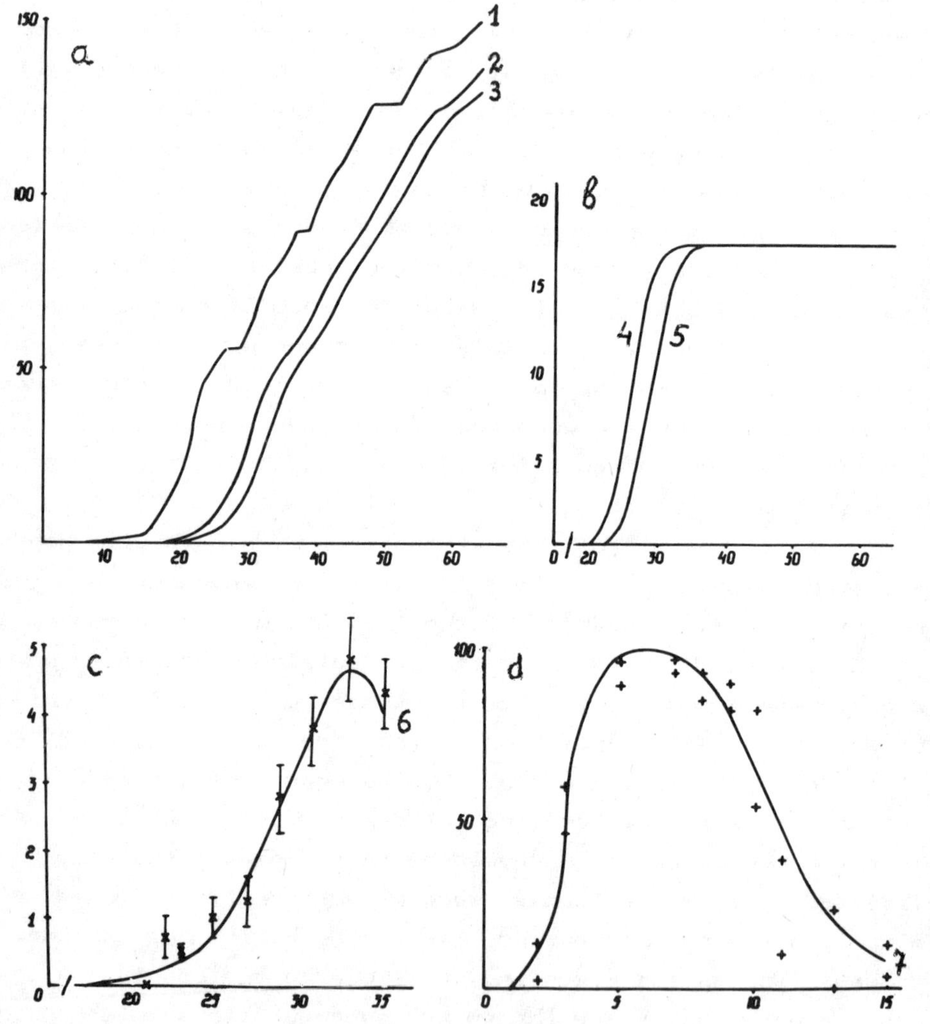

Figure 12. FLM plot for hepatocytes of the regenerating liver of the August strain rats
1-$q_S(t)$;2-$q_{G_2}(t)$;3-$q_M(t)$;4-$q^*_{G_2}(t)$;5-$q^*_M(t)$;6-$I_M(t)$;7-FLM(t);x-experimental $I_M(t)$ values; + - experimental FLM(t) values. On the abscissa-time (hr):a,b,c after PHE, d-after pulse labelling.

We now turn to the analysis of data obtained by Fabrikant in the study of liver regeneration in the August strain rats. The data includes experimental FLM(t), thus providing a wider scope for employing special procedures of kinetic analysis of induced cell proliferation. Fig.12 illustrates the main stages in plotting $I_M(t)$ and FLM(t) curves using those data. Estimates of temporal parameters of the S-, G_2- and M-phases were sought by fitting theoretical $I_M(t)$ and FLM(t) curves to their experimental values by the least-squares technique. The Nelder-Mead algorithm described in Himmelblau's book [20] was used in searching the minimum of the functional. The following estimates were obtained(hr): $\bar{\tau}_S$=8.63; σ_S=2.09; $\bar{\tau}_{G_2}$=2.87; σ_{G_2}=0.96; $\bar{\tau}_M$=1.01; σ_M= 0.29. Investigation into the sensitivity of FLM(t) computation algorithm has corroborated Macdonald's conclusion [24] to the effect that it is difficult to obtain from FLM information on the value of mitosis length variation coefficient. The function $I_M(t)$ is also characterized by low sensitivity to variations in the value of σ_M.

During the first 40 hr after PHE the index $q_S(t)$ shows the same regularities as those described in the foregoing for the regenerating liver of the MacCollum strain rats. The existence of the initial fraction of hepatocytes (equal in that case to about 57% of $n_\Sigma(0)$) entering synchronously the mitotic cycle (provided the whole prereplicative period is included in the structure of the first mitotic cycle following PHE) is evidenced by a distinct plateau in the $q_S(t)$ index in the interval of 26-30 hr after the operation. To describe the $q_S(t)$ index on the interval up to 29 hr formula (1) was used.

In approximating $F_{PS}(t)$ by Γ-distribution with the parameters α and β, the following estimates of parameters were obtained as a result of optimization: α_{PS}=49.006±0.004, β_{PS}=2.285± 0.010;C_0=0.567. Thus, satisfactory estimates of the temporal parameters of the prereplicative period are: $\bar{\tau}_{PS}$=21.44 hr and σ_{PS}=3.06 hr. There are but minor differences in the numerical values (with a possible exception of mitosis) of the first mitotic cycle of hepatocytes following PHE for the August and MacCollum strain rats investigated by Fabrikant.

Following from the asymptotic properties of the index q(t) is the equality

$$\frac{1}{\bar{\tau}_S}\int_0^\infty \tilde{I}_S(t)dt=\frac{1}{\bar{\tau}_M}\int_0^\infty \tilde{I}_M(t)dt,$$

hence

$$\frac{\bar{\tau}_S}{\bar{\tau}_M} = \frac{\int_0^\infty \tilde{I}_S(t)dt}{\int_0^\infty \tilde{I}_M(t)dt} \quad . \qquad (2)$$

These relations may also be used for rough estimation of the temporal parameters of the mitotic cycle in systems with induced DNA synthesis maintaining balance of cell entering the S- and M-phases, i.e. where all DNA sythesizing cells invariably undergo mitotic division. The estimates of the $\bar{\tau}_S$ and $\bar{\tau}_M$ parameters obtained in the foregoing are not at variance with equation (2).

Comparison of the results of kinetic analysis of liver regeneration in the August strain rats with experimental data on rats of some other strains obtained by using the method of continuous ^{3}H-thymidine labelling [16,30,35] shows that hepatocytes enter the first cycle in two large portions: the first ($\simeq$ 57%) appears to be involved in proliferation processes directly in response to PHE, while the second ($\simeq$ 35%) follows with a significant delay. This conclusion is based on the assumption that the durations of prereplicative periods for both fractions are similarly distributed random variables. Otherwise the case in question would be that of the existence of cell subpopulations differing widely in their prereplicative period durations, i.e. of bimodal $F_{PS}(t)$ distribution, rather than of different moments when separate fractions of a hepatocyte population enter the prereplicative period. It is expedient to single out specifically situations when there is good reason to believe that the duration of a prereplicative period (or of its part) rises or drops, i.e. the principal intracellular processes of preparation for DNA replication are accelerating or slowing down. Such situations should be distinguished from cases when for some cause there is a delay in the entry of a fraction of cells into the reproductive cycle. In the meantime, a delay may occur not only prior to the entry of cells into the transformation period but also in the course of transition from the transformation period into the G_1-phase of the mitotic cycle. Non-random, i.e. equal for all cells of the fraction, time of delay does not reduce the degree of synchronism in entering the S-phase, which may serve as a guide in choosing one of the two ways of interpreting experimental data.

For the sake of comparison let us consider the results of kinetic analysis of the data reported in reference [10] and obtained in investigating a confluent culture of human lung fibroblasts stimulated to proliferation. The pattern of the $q_S(t)$ and $P_S(t)$ curves (Fig.13) plotted from those experimental data clearly shows the presence of about 60% of cells (fraction A) which by 20 hr after stimulation of the culture start to synthetize DNA for the first time. It is only after 30 hr

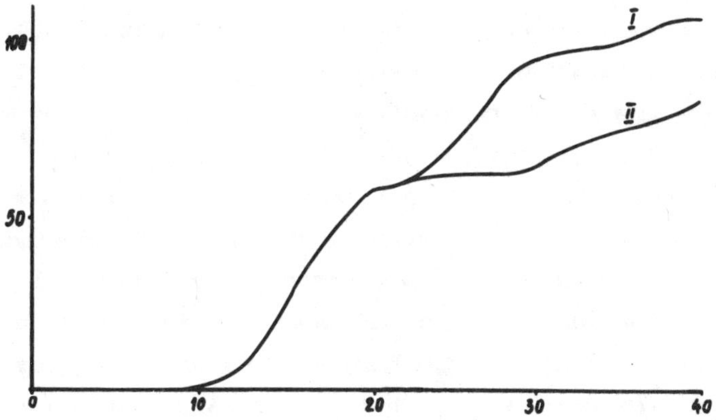

Figure 13. The results of kinetic analysis of data from [10]. 1- $q_S(t)$,2- $P_S(t)$.

following stimulation that the other portion of cells (fraction B) begins transition to initial DNA synthesis, yet far less synchronized than in the case of fraction A. From 20 hr to 30 hr the only processes that are taking place are those of the re - involvement of fraction A cells in the mitotic cycle. However, far from all the cells of the fraction can accomplish repeated DNA sythesis even before the end of the observation period. Two ways of interpreting these data are possible. It may be assumed that the transformation period of fraction B cells begins much later than that of fraction A cells, and the time of delay of fraction B at rest is not included in the actual (random) duration of the processes of preparation for DNA synthesis. This interpretation also implies that fraction B cells enter, after a certain delay, the transformation period less synchronously than fraction A cells. On the other hand, it is conceivable that, although the proliferative stimulus is received by all the cells of a population, the processes of the prereplicative period in fraction B cells are taking place at a much slower rate than in fraction A cells, and the longer mean duration of the prereplicative period for fraction B cells (without an accompanying sharp drop in variability) accounts for their less synchronized transition to DNA sythesis. The latter interpretation seems to us more adequate.

The data reported by Fabrikant [14] permit independent study of $q_S(t)$ patterns for hepatocytes from different zones of the hepatic lobule. Such an investigation has demonstrated that cells of the "rapid" synchronous fraction are positioned closer to the periphery of the hepatic lobule, while the "slow" fraction is located primarily in the central zone. The periportal zone and the peripheral half of the middle zone also account for the vast majority of re-entries of cells into the mitotic cycle, whereas the bulk of proliferating hepatocytes of the central zone undergo but one division within 64 hr of observation. This conclusion is supported by analysis of FLM curves obtained for different cell positions inside the hepatic lobule [29,30]. An important point is that the plateaus in $q_S(t)$ curves for different hepatic lobule zones are related to the same time interval following PHE and always coincide in time with the plateau in the summation curve presented in Fig.12. This observation indicates that the strategy managing the processes of cell entry into and exit from the cycle includes as an essential element setting up a "reserve" of cells outside the cycle and dynamic redistribution of cells between the "reserve" and the state of active proliferation. Meanwhile the transition of cells to proliferation proceeds by synhronized entry of certain groups of cells into the prereplicative period or directly into the G_1-phase of the mitotic cycle.

Let us present the above results within the framework of a general kinetic scheme to be devised on the basis of the presence of three portions in the $q_S(t)$ curve plotted from experimental data (Fig.12), the first corresponding to the time interval of 0 to 29 hr, the second - from 29 to 39 hr and the third - from 39 to 52 hr. The available experimental evidence on continuous introduction of ^{3}H-thymidine [16,29,30,35] and the results of our analysis of Fabrikant's data [14] on distribution of pulse labelled and dividing cells among different hepatic lobule zones suggest that the first portion of the $q_S(t)$ curve represents the initial entry of the first part of the original hepatocyte population (denoted for convenience as fraction A) into the DNA synthesis phase, the second portion of the $q_S(t)$ curve represents mainly re-entry of some of fraction A cells into the next cycle, and the third portion relates to the initial entry into the S-phase of the rest of the cells of the original population induced to synthesize DNA (fraction B). According to our estimates, fraction A accounts for about 57% and fraction B for approximately 35% of the hepatocyte population remaining in the liver after PHE. These estimates may prove somewhat too high since a certain contribution of re-dividing cells to the $q_S(t)$ function increment in the first and third portions cannot be ruled out. A part of the original hepatocyte population equal to about 5-10% $n_\Sigma(0)$ does not appear to start synthesizing DNA before the end of the observation period (64 hr after PHE).

The first $\simeq$ 15-16% $n_\Sigma(0)$ of hepatocytes of fraction A enter mitosis already at 30 hr after PHE. Those cells with their descendants (i.e. a total of about 30% $n_\Sigma(0)$) at once enter the G_1-phase of the second mitotic cycle. The next 17-18% $n_\Sigma(0)$ pass through the stage of mitosis in the period of 30 to 34 hr, increasing their numbers as a result of mitotic divisions to $\simeq$ 35% $n_\Sigma(0)$, yet without involvement into the second cycle at least before 46-48 hr after PHE. Fraction B cells do not enter the G_1-phase of the first mitotic cycle before the equivalent number of new cells are formed (as a result of division of fraction A cells). As shown in reference [48] such interrelations between individual fractions are nonexistent in a stationary cell culture stimulated to DNA synthesis.

Most likely the period of transformation for fraction B cells corresponds to early stages of liver regeneration and its completion does not necessarily entail the transition of a cell to the G_1-phase of the first mitotic cycle following PHE. As regards the duration of the transformation period for fraction B hepatocytes it may only be hypothesized that its maximum value is much below 34-35 hr since otherwise one would expect a considerably less synchronized entry of fraction B into the period of DNA synthesis in the interval from 39 to 48 hr after PHE. In other words, the time fraction B cells are delayed at the transformation period/ G_1-phase boundary may be regarded as non-random and sufficiently long to ensure completion of the transformation of all fraction B cells; acting as the stimulant for their subsequent synchronized transition to the G_1-phase of the mitotic cycle is, presumably, accumulation of an equivalent number of new hepatocytes in the population. These arguments sugges t that the processes of hepatocyte transition to the period of transformation and thence to the G_1-phase are controlled by different mechanisms. Consequently, the scheme presented by Yakovlev et al.[50] is supplemented with the proposition that different temporal prereplicative period structures may correspond to differing groups of cells (Fig.14). It is evident that the regularities in the pattern of hepatocyte proliferation processes following standard PHE as described above do not fit in with Fabrikant's concepts outlined at the beginning of this section, and they are also at variance with the Smith and Martin model (see Chapter II) claiming universality.

Kinetic analysis has revealed a coordinated behaviour of individual hepatocyte fractions in the course of active liver regeneration resulting in a renewing subpopulation of cells which, at each instant of time, are outside the mitotic cycle. That phenomenon may be designated as "dynamic replacement of cells". Experimental findings reviewed in the next section indicate that it is this subpopulation of cells, its renewal conforming to the above regularities, that ensures performance of specialized functions of the regenerating liver.

The kinetic regularities of induced cell proliferation are expressed in terms of different hepatocyte subpopulation (fraction) sizes, without regard for the processes by hypertrophy

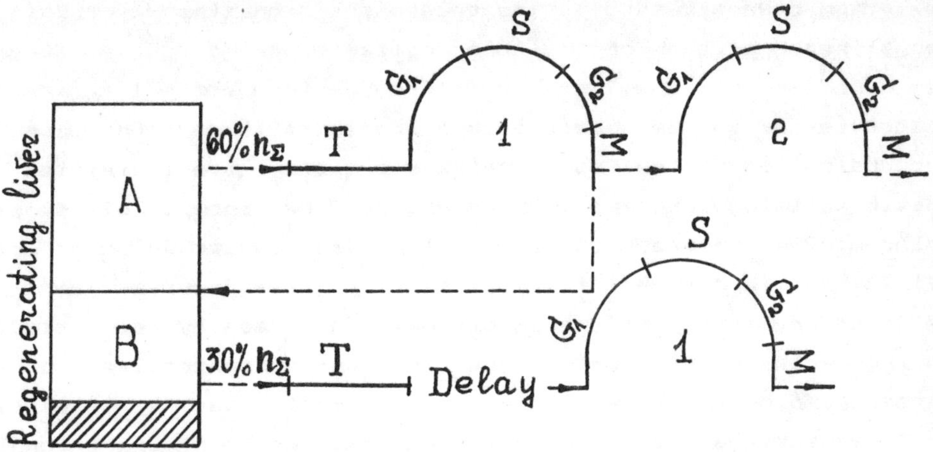

Figure 14. Schematic representation of hepatocyte proliferation kinetics at the early stages of the rat liver regeneration.

and polyploidization of cells. Certain known mechanisms of polyploidization associated with peculiarities inherent in the division of mononuclear and binuclear hepatocytes create serious complications in the analysis of radioautographic evidence. The main handicap is due to the fact that mitosis of certain cells may culminate in the formation of a descendant of superior ploidy which leads to an unaccounted rise in the q_S index and a drop in P_S. However, according to Gerhard's model [18] the fraction of such cells in the remnant of the liver after PHE amounts to about 18%. The contribution made by that fraction in the calculation of the q_S and P_S indices is counteracted by the mean mitotic duration estimate, and therefore further correction of kinetic analysis results seems unnecessary. It is impracticable at present to determine precisely the representation of hepatocytes of different ploidy in the composition of the dynamic cell reserve at all the stages of the "compensatory" period of the liver growth. Further investigation of the problem may result in a more sophisticated interpretation of dynamic replacement which would take into account the functional inequality of cells differing in ploidy [6].

5.3. Dynamic Replacement of Hepatocytes, a Mechanism Maintaining Specialized Functions of the Regenerating Liver

In this section we shall discuss in general terms the problem of coordination of the functional and proliferative activities of hepatic cells and attempt to substantiate the concept of dynamic replacement of hepatocytes as a cell population mechanism ensuring specialized functions of the liver in the course of its regeneration. There are comparatively few, mostly old studies dealing with specialized functions of hepatocytes during regeneration of the liver. This is, presumably, due to the fact that the regenerating liver of the mammals has been traditionally used as an experimental model to investigate biomolecular and biochemical mechanisms of cell reproduction. In such studies diverse metabolic shifts seen after PHE are almost invariably related to events occurring in either dividing cells or in those committed to division. However, as rightly stressed by Rabes [30], a parenchymal cell population of the regenerating liver after PHE faces a twofold problem of maintaining the major functions of the organ at a certain nominal level, and ensuring, at the same time, restoration of its initial mass.

Available experimental evidence suggests that at least some of the specialized functions of the liver are not depressed during regeneration [3,4,17,39], the urea production function providing a most graphic example. Reference [17] demonstrates that there is no accumulation of ammonia nitrogen in the blood of animals undergoing PHE, while there is a sharp rise in the urea production on a per-hepatocyte basis. The authors believe that acceleration of the passage of metabolites through the urea cycle is essential for a rise in the ornithine levels in hepatic cells which, with attendant increase in ornithine decarboxylase activity, ensures accumulation of polyamines and, consequently, stimulation of RNA synthesis. Contrary to the authors' view it may be presumed that it is hepatic cells quiescent with respect to proliferation processes that are responsible for higher urea production. As regards cells about to begin DNA synthesis, the urea cycle, on the contrary, goes off due to increase in ornithine decarboxylase activity. With accelerated formation of pyrimidine nucleotides such an increase may occur under the influence of carbamylphosphate and dihydrorothic acid.

It goes without saying that such explanations are of purely speculative character both because of the intricate biological scheme and the impracticability of determining precisely what changes in ornithine decarboxylase activity are taking place in just fraction A cells. For instance, it is quite possible that the activity of the enzyme in those cells is even decreasing. This possibility is evidenced by data from [27] obtained in studying reactions of hemopoietic tissue to perturbations of polyamine synthesis. In that work it is shown, in particular, that following administration of an agent, which selectively and irreversibly inhibits ornithine decarboxylase, a two to four fold increase was found in the number of hemopoietic precursor cells.

The functional load on the urea cycle following PHE appears to be quite formidable, requiring mobilization of all the resources of the hepatic tissue remaining after the operation. Indeed , administration of ammonium acetate or a complete mixture of amino acids to animals (simultaneously with PHE) caused a more than six-fold increase in the blood levels of ammonia nitrogen; at the same time sham-operated animals showed only a slight elevation of short duration [17].

In studying various functional loads on the mammalian liver, two levels of adaptation to them can be distinguished. One comprises metabolic shifts aimed at enhancing the functional activity of each hepatocyte competent for a given function, and the otherrepresents induction of cell proliferation (or, in some cases, hypertrophy). With minor loads the former level of adaptation prevails, but when the load exceeds its potentialities need will arise for a larger total number of functioning elements, causing the transition of the cell system to the latter adaptation level. Quite often this transition proceeds against the background of a pronounced decompensation of cell functions.

Below are some examples illustrating proliferative reactions of hepatic cells under functional loads. Review [1] presents experimental findings indicating an intimate relationship between induction of microsomal oxidation enzymes and proliferative reactions of hepatic cells. Specifically, it was shown that administration of phenobarbital and 3-methylcholanthrene may stimulate cell reproduction in the liver of young rats. The liver of adult animals reacts to a similar load by hypertrophy of hepatocytes. Protein-rich diet both gives rise to induction of ornithinetranscarbamylase (and, most probably, of other urea cycle enzymes) and triggers cell proliferation. An interesting version of the experiment was that in which preliminary treatment with phenobarbital and attainment maximum possible intensity of the ensuing proliferative response were followed by administration of 3-methylcholanthrene producing an additive effect. The additive effect also resulted from a combined action of protein-rich diet and phenobarbital.

Experimental galactosemia likewise stimulates proliferation of hepatocytes in the intact liver [21]. On the 2nd-4th day of feeding rats with galactose-rich food there occurs activation of galactoso-I-phosphate-uridyl-transferase, an enzyme involved in the transformation of galactose into glucose. In the period from the 5th to the 13th day the enzyme activity diminishes, despite the continuing galactose-rich diet, and starting on the 12th day induction of hepatocyte proliferation is observed.

Prolonged suppression of the protein-synthesizing function of the liver with cycloheximide also induces replicative DNA synthesis [38]. Needless to say there are different biochemical schemes accounting for the way in which modification of a certain metabolic or biosynthetic route may lead to induction of proliferation processes in the cell. It is from the standpoint of such schemes that the authors interpret the data described above. However it appears reasonable to us to see in the regularities of hepatocyte proliferation response to a functional load an important evidence of the existence of a cell-population mechanism integrating proliferation processes and specific functions of hepatocytes. The distincture feature of such a mechanism is that while <u>in some cells</u> of the population maximum adaptation of metabolism to a functional load is taking place, <u>in the others</u> processes of preparation for DNA synthesis and mitotic division are being induced. For the sake of a further discussion of this viewpoint let us consider some results of studying the effect of functional loads on the process of liver regeneration after PHE.

Whenever PHE is performed in the setting of artificial intensification of a hepatic function, there is a temporary inhibition of the processes of induced hepatocyte proliferation. For instance, a significant suppression of mitotic activity recorded at 30 hr after PHE was noted when the animals were administered (according to a certain scheme) large doses of glucose or fructose, as well as of glucose with insulin [37]. In addition, it is noteworthy that preliminary administration (6 hr prior to PHE) of saccharose suppressed nearly completely DNA synthesis in the rat liver at 26 hr after the operation [34].

Van Cantfort and Barbason [42] conducted a comparative study on the dynamics of the mitotic index and the activity of cholesterol-7α-hydroxylase (the key enzyme in the synthesis of bile acids) after PHE at different times of the day. Cholesterol-7α-hydroxylase molecules are short-lived, their half-life period is about 4 hr and, therefore, the marked diurnal variations in the activity of the enzyme in the normal liver are due mainly to alterations in the rate of its synthesis. In the experiments of Van Cantfort and Barbason the following regularities were noted. If PHE was performed in the morning (when the activity of cholesterol-7α-hydroxylase in the liver is at its minimum),the enzyme activity was falling off and started to rise only following the first wave of mitoses, after which the diurnal rhythm of enzymatic activity was gradually approaching the normal level. In the experiment thus designed, within the period of observation (84 hr after PHE), in addition to the first principal wave of mitoses there were two more waves of lower intensity which were almost in antiphase with variations in the activity of cholesterol-7α-hydroxylase.When the operation was performed in the evening (the time of maximum enzyme activity) the activity of cholesterol-7α-hydroxylase kept on following its diurnal pattern, being no different from the control values yet, on reaching the minimum it remained low throughout 24 hr. Then followed the first wave of mitoses, much more intensive than that observed after morning PHE, and subsequent sharp rise in enzyme activity.The regular character of diurnal variations in enzyme activity was practically restored only after the second wave of mitotic divisions. The prominent feature of hepatocyte proliferation reaction in rats operated upon in the setting of enhanced hepatic function (with cholesterol-7α-hydroxylase as the marker) is that, though the first wave of mitoses in that case is delayed, its intensity is higher than in rats operated upon at a time of reduced hepatic function. This character of adaptive shifts in a cell system lacking a clear-cut space distinction between proliferation processes and specific tissue function of cells indicates that effective coordination of these processes may be attained through their separation in time. The results of Van Cantfort and Barbason clearly demonstrate a close interrelation between restoration of functional activity of the liver after PHE and changes in the size of a hepatocyte population.

In a later work [2] Barbason et al. undertook a profound investigation into the hormonal regulation of the diurnal rhythm of hepatocyte proliferation in young rats. Unfortunately, that work contains no experimental data on hormonal effects upon specialized hepatic functions (cholesterol-7α-hydroxylase activity) which could promote better understanding of the cell-population mechanisms regulating them.

Another study [8] has revealed a similar relationship between the character of proliferative response of hepatocytes (estimated by the dynamics of the inclusion of ^{131}J-deoxyuridine and mitotic index) and the time PHE is performed. Administration of phenobarbital immediately after PHE induced noticeable changes in cell kinetics (delay in the transition of cells to DNA synthesis and its subsequent intensification) when the animals were operated upon in the evening.

Specialized functions of hepatocytes exhibit different priorities in relation to proliferation processes. High-priority functions do not fall off in the course of liver regeneration, and their artificial reinforcement may cause profound and prolonged suppression of hepatocyte proliferation induced by PHE. It is conceivable that such functions are characterized by competition for common substrates with biochemical reactions involved in the preparation of a cell for DNA synthesis. Specifically, manifestations of such competition should be expected with a load on the urea production function of hepatocytes at the initial stages of liver regeneration. Competitive interrelations of cell functions are discussed in detail in a review by Brodsky and Uryvaeva [5].

The existence of specialized functions that have absolute priority as regards proliferation is indicated indirectly by the results of PHE experiments differing in extent. Resection of a portion of the liver accounting for 9 to 34% of the initial mass brings about only a minor intensification of DNA synthesis; removal of 43% of the liver tissue leads to a sharp rise in DNA synthesis [7]. Maximum intensity of induced hepatocyte proliferation appears to occur in the case of the standard operation ablating 2/3 of the liver. A more extensive resection (82%) of the liver results in a considerable delay in the entry of cells into the periods of DNA synthesis and mitosis [44,45], while with 90% PHE there is no induction of cell proliferation altogether, at least within 40 hr after the operation, and the animals often die under such conditions [40]. One of the possible explanations for the profound suppression of proliferative reaction in the case of subtotal hepatectomy is that an intensive load on the major (priority) functions of the liver requires enhanced functional activity of all remaining hepatocytes which, for that reason, are not involved in the process of regeneration of hepatic parenchyma. A similar situation is possible in the case of toxic effects. Intoxication of the organism with carbon tetrachloride reduces the fraction of cells entering DNA synthesis in response to PHE [36].

Analysis of the above experimental evidence urges that attention must be focussed again on the well-known proposition of antagonism between proliferation processes and specific functions of cells. Without going into a discussion with investigators who refer in their works to different exceptions to this fundamental theoretical principle of present-day biology, we shall confine ourselves to giving an example clearly demonstrating that a hepatocyte involved in proliferation is incapable of discharging the whole range of its differentiated functions. Uryvaeva and Faktor [41] have shown that in the interval from 50 to 60 hr after PHE centrolobular hepatocytes of the mouse liver temporarily lose sensitivity to the toxic effect of carbon tetrachloride. It will be recalled that it is at that time after the operation that induced transition of B-fraction cells to the mitotic cycle should be expected. CCl_4 is known to undergo in hepatic cells microsomal oxidation involving cytochrome P-450, and injury to centrolobular hepatocytes is due to their active detoxication function. Thus, reprogramming of the type of metabolism in a proliferating cell leads to a temporary loss of its specific functional competence. Therefore, the dynamic reserve of hepatocytes serves to maintain the functional activity of the liver while the proliferating cells ensure restoration of the organ's mass.

The problem of how the small fraction (1/3) of parenchymal cells remaining after the operation ensures simultaneously both performance of the key functions of the liver and the required increase in the size of the cell population will be viewed in a new light once we take into account the peculiarities of kinetic behaviour of hepatocytes following PHE. The results of kinetic analysis presented in the preceding section suggest a concept that coordination of specialized functions and cell multiplication in the regenerating liver is accomplished at the cell-population level owing to the non-simultaneous involvement of individual fractions of the hepatocyte population in the processes of proliferation and redistribution of functional activity between the fractions. At the initial stages of liver regeneration the principal functional load is accounted for by fraction B cells.However, so small a population of cells (≃10% of the total number of parenchymal cells in the intact liver) may effectively ensure performance of priority functions of the liver only within a limited period of time. Therefore, the reserve of cells that is built up by certain time after the operation (as a result of fraction A cells multiplication) undertakes an increased functional load, while fraction B is involved in proliferation processes. It is not known whether immediately after mitotic division the reserve cells have the competence required for specialized functions or whether certain time is still required for its acquisition during which temporary decompensation of some hepatic functions may be tolerated.

Also remaining unknown is the specific nature of the system controlling redistribution of the functional duties of hepatocytes. An important factor in the formation of the system is the anatomic structure of the hepatic lobule since, for example, specialized biosynthetic processes in the intact liver are confined mainly to centrolobular hepatocytes [29].Compelling evidence of the existence of a cell-population system controlling integration of hepatocyte functional and proliferative activities can be found in the absence of the dynamic cell replacement phenomenon under the in vitro conditions [47,48].

The temporary blocking of fraction B cells prior to their entry into the G_1-phase may be accounted for by more pronounced catabolic processes in those cells as compared to fraction A cells. Indeed, the data reported in reference [28] indicate that at the initial stages of liver regeneration fraction B cells accumulate acid phosphatase-containing autophagocytic vacuoles. A number of experiments on cell cultures have shown that autophagocytosis is not an essential element of the mechanism that activates nuclear chromatin at the time a cell passes from quiescence to active proliferation (see [28]). It seems that the most plausible view is that autophagocytosis serves as an "emergency" route of energy metabolism used by a cell as a mechanism of adaptation to functional loads. In this connection an important role is, apparently, played by the hormonal system of carbohydrate metabolism regulation which reacts to drastic depletion of glycogen supplies in the organism following partial hepatectomy. In this context it is interesting to note that the relative intensity of vacuole formation in response to glucagon administration is considerably lower in the regenerating liver as compared to the intact organ [43].

5.4. A Simple Mathematical Model of Liver Response to Partial Hepatectomy of Different Extent

A full mathematical description of dynamic cell replacement in the regenerating liver is quite an arduous task. It is possible, however, to consider the simplest version of the model reflecting only one aspect of the phenomenon under review. The model is formulated in terms of reliability theory for redundancy devices and is based on the following simplifying assumptions:

1. Let m be the number of cells remaining in the liver after

PHE, and the fraction of the resected part of hepatic tissue (in relation to the intact liver) be equal to ρ. Thus, prior to PHE the number of cells in the liver was $m/(1-\rho)$. Immediately after the operation three groups of cells differing in further kinetic behaviour are formed in the liver remnant: group 1 consists of m_1 cells totally excluded from proliferative reaction and destined to ensure specialized functions of the liver, group 2 comprises m_2 cells which undergo a single division, and group 3 is made up of m_3 cells which pass twice through the mitotic cycle.

2. Group 1 cells may be in one of the two alternative states: the state of normal functioning or that of failure, and the failing cells are never repaired. The total functional load is uniformly distributed among group 1 elements, but the failures necessitable redistribution of the load among those cells of the group that remain operative. The rate of failures λ depends on the number of failing elements i in the following manner

$$\lambda_i=\lambda_n\left[\frac{m}{(1-\rho)(m_1-i)}\right]^{1-\nu},\quad \nu\neq 0\ ,\quad m=m_1+m_2+m_3\ ,$$

where λ_n is the nominal failure rate for the intact liver and ν is the dependence form coefficient. As shown in reference [31] the probability of i failures by the moment t may be expressed by the formula

$$P_i(t)=\prod_{j=0}^{i-1}(m_1-j)^{\nu}\sum_{j=0}^{i}\frac{\exp\left[-(m_1-j)^{\nu}\lambda_n\left(\frac{m}{1-\rho}\right)^{1-\nu}t\right]}{\prod_{\substack{l=0\\ j\neq l}}^{i}\left[(m_1-l)^{\nu}-(m_1-j)^{\nu}\right]}\ ,$$

$$\nu\neq 0\ ,\ i=1,2,\ldots,m_1-1\ ;$$

$$P_0(t)=\exp\left[-m_1^{\nu}\lambda_n\left(\frac{m}{1-\rho}\right)^{1-\nu}t\right].$$

3. Groups of cells 2 and 3 synchronously enter the mitotic cycle immediately after PHE ($t=0$). Cells of group 3 are involved in the second mitotic cycle only after all m_3 cells of the group

complete the first division. The entry of group 3 cells into the second cycle is also assumed to be synchronous. The durations of the 1st and 2nd mitotic cycles obey uniform distribution, the mean duration of the 1st cycle is longer than of the 2nd. All cells of groups 2 and 3 that have undergone one or two divisions, respectively, pass into group 4 and acquire the ability to fulfill the same functions as group 1 cells. Cells belonging to group 4 are not prone to failure.

4. Let us denote group 1 cells not failing by the moment t by $A(t)$ and the number of cells existing in group 4 at the moment t - by $B(t)$. Failure of the entire cell system is assumed to be the event $A(t)+B(t)<\gamma$ where γ is the critical number of cells capable of ensuring the nominal level of hepatic functions. The probability of failure of the cell system at the moment t is introduced:

$$Q(t) = \mathbb{P}\{A(t) + B(t)<\gamma\} ,$$

which, with the assumptions made, may be expressed in analytical form. The functional $J[Q(t)]$ may serve as reliability criterion for the system. Fixing the observation period $[0,T]$, we shall choose as such a functional the norm $\|Q(t)\|$ in the class $L_1(0,T)$. The value T exceeds the total duration of two mitotic cycles. Our interest is only with the relationship between J and the vector of the parameters m_1, m_2 and m_3, as the values of the other parameters of the model assumed to be fixed. Thus, as regards its reliability, the optimality of the system under consideration will be determined by the condition:

$$\min_{m_1,m_2,m_3} J(m_1,m_2,m_3;T), \quad m_1+m_2+m_3=m .$$

Investigation of the model behaviour on a computer at certain parameter values has resulted in the following main conclusions:

1. If the selected variable γ is considerably smaller than m, and the magnitude of the functional load on the liver remnant (characterized by the parameter ρ) is increasing, then, starting from a certain ρ value, the condition of optimal cell system reliability is ensured by stimulation of cell division. In other words, with an increase in the resected portion of the liver

within such limits when the number of cells remaining in the liver is well in excess of the critical level, the processes of cell proliferation are growing in importance for ensuring the functional reliability of a hepatocyte population.

2. If, on the contrary, the variable γ is close to m, i.e. the number of cells remaining in the liver after the operation exceeds but slightly the critical value, then maximum reliability will be ensured provided all the cells are performing specialized functions without being involved in proliferation processes.

These conclusions may provide one of the possible explanations for the relationship between liver proliferative response and PHE extent observed in experiments and discussed in the foregoing (Section 5.3). The results of studying the properties of the model correlated with the fact of a drastic suppression of the proliferative reaction of the liver after subtotal hepatectomy suggest that the critical number of the rat hepatocytes capable of ensuring the nominal level of the priority specialized functions of the liver at early stages of its regeneration accounts for about 10% of the size of the parenchymal cell population in the intact liver.

The final formulation and formalization of the principles of dynamic replacement of cells are of importance not only as regards the problem of regeneration and pathology of the liver; these principles also offer much promise for the realization of bionic approach to the problems involved in the analysis and synthesis of high-reliability complex technical systems.

REFERENCES

1. Argiris, T.S. Stimulators, enzyme induction and the control of liver growth, In: Control of Proliferation in Animal Cells, Gold Spring Harbor Laboratory, 46-66, 1974.
2. Barbason,H., Herens, Ch., Mormont, M.-C. and Bounzahzah C. Circadian synchronization of hepatocyte proliferation in young rats: the role played by adrenal hormones, Cell Tiss. Kinet,20, 57-67, 1987.
3. Becker, F.F. Structural and functional correlation in regenerating liver, In: Biochemistry of Cell Division, 113-118,1969.
4. Becker, F.F. and Lane, B.P. Regeneration of the mamalian liver. VI.Retention of phenobarbital-induced cytoplasmic alterations in dividing hepatocytes, Amer.J.Pathol., 52, 211-221, 1968.
5. Brodsky,W.Ya. and Uryvaeva.I.V.Cell polyploidy and its relation

to tissue growth and function,Internat.Rev.Cytol.,50,275-323, 1977.

6. Brodsky, W.Ya. and Uryvaeva, I.V. Cell polyploidy.Proliferation and differentiation, Nauka, Moscow, 1981 (In Russian).
7. Bucher, N.L.R. and Swaffield, M.N. The rate of incorporation of labelled thymidine into the deoxyribonucleic acid of regeneration rat liver in relation to the amount of liver excised, Cancer Res., 24, 1611-1625, 1964.
8. Bürki, K., Schindler, R. and Ptenninger, N. Studies on liver regeneration. II. Effect of phenobarbital on the onset and pattern of rat liver regeneration following partial hepatectomy.Cell Tiss.Kinet., 4, 529-538, 1971.
9 .Chistopolsky, A.S. and Zimin. Yu.P. Mean mitotic duration of hematopoietic tissue cells in rats, Cytology,19,824-829,1977 (In Russian).
10.Ellem. K.A.O. and Mironescu, S.The mechanism of regulation of fibroblastic cell replication. I. Properties of the system, J.Cell.Physiol., 79, 389-406, 1972.
11.Fabrikant, J.I. Cell proliferation in the regenerating liver and the effect of prior continuous irradiation, Radiat.Res., 32, 804-826, 1967.
12.Fabrikant, J.I. Kinetic analysis of hepatic regeneration, Growth, 31, 311-315, 1967.
13.Fabrikant, J.I. Rate of cell proliferation in the regenerating liver, Brit.J.Radiol.,41,71, 1968.
14.Fabrikant, J.I. The kinetics of cellular proliferation in regenerating liver,J.Cell Biol., 36, 551-565,1968.
15.Fabrikant, J.I. Radiation response in relation to the cell cycle in vivo, Amer.J.Roentgen.,Radium Ther., Nucl.Med.,105, 1969.
16.Fabrikant,J.I. Size of proliferating pools in regeneraring liver, Exper.Cell Res., 55, 277-279, 1969.
17.Fausto, N., Brandt, J.T. and Kesner, L. Interrelationships between the urea cycle, pyrimidine and polyamine synthesis during liver regeneration, In: Liver Regeneration after Experimental Injury, Stratton Intercontinental Medical Book Corp., 78-93, 1975.
18.Gerhard, H. A quantitative model of cellular regeneration in rat liver after partial hepatectomy, In: Liver Regeneration After Experimental Injury, Stratton Intercontinental Medical Book Corp., 340-346, 1975.
19.Hartmann, N.R., Gilbert, C.M., Jansson, B., Macdonald, P.D.M., Steel, G.G, and Valleron, A.J. A comparison of computer methods for the analysis of fraction labelled mitoses curves, Cell Tiss.Kinet., 8, 119-124, 1975.
20.Himmelblau, D.M. Applied nonlinear programming, McGraw-Hill Book Company, 1972.
21.Kostyrev, O.A. and Solovyeva, N.A. DNA synthesis and mitoses stimulation in the rat liver in experimental galactosemia induced by prolonged administration of galactose,Cytology, 17,1042-1046, 1975 (In Russian).
22.Kubitschek, H.E. Normal distribution of cell generation rate, Exper. Cell Res., 26, 439-450, 1962.
23.Lindgren, A.L., Riley, E.F. Use of the wound response to demonstrate latent radiation damage and recovery in X-irradiated inte-mitotic cells of the rat lens epithelium, Radiat.Res., 54, 411-430, 1973.

24.Macdonald, P.D.M. Statistical inference from the fraction labelled mitoses curve, Biometrika, 57, 489-503, 1970.
25.Marshall, W.H., Valentine, F.T. and Lawrence, H.S. Cellular immunity in vitro. Clonal proliferation of antigen - stimulated lymphocytes, J.Exper.Med., 130, 327-342, 1969.
26.Natchwey, D.S., Cameron, I.L. Chapter 10, In: Methods in Cell Physiology, Academic Press, New York, 3, 213-259, 1968.
27.O'Conor,G.T., McCann, P.P., Wharton III, W.W. and Niskanen,E. Haematological cell proliferation and differentiation responses to perturbations of polyamine biosythesis, Cell Tiss.Kinet., 19, 539-546, 1986.
28.Petrovichev, N.N. and Yakovlev, A.Yu. Some peculiarities of the process of autophagocytic vacuole formation in the regenerating rat liver,Cytology,17,1087-1089,1975(In Russian)
29.Rabes, H.M. Kinetics of hepatocellular proliferation after partial resection of the liver,In:Progress in Liver Diseases, 83-89, 1976.
30.Rabes, H.M., Wirsching,R., Tuczek, H.V. and Iseler, G. Analysis of cell cycle compartments of hepatocytes after partial hepatectomy, Cell Tiss.Kinet., 9, 517-532, 1976.
31.Raikin.A.A. Probabilistic models of redundant devices,Nauka, Moscow, 1971 (In Russian).
32.Rao,P.N. and Engelberg, J. Mitotic duration and its variability in relation to temperature in HeLa cells,Exper.Cell Res.,52, 198-208, 1968.
33.Rixon,R.H. and Whitfield,J.E.The control of liver regeneration by parathyroid hormone and calcium,J.Cell Physiol., 87, 147-155, 1976.
34.Sapigni,T. and Melandri,P. Effetto della vacuolizazione da saccarosio sulla rigenerazione epatica nel ratto,Boll.Soc. Ital.Biol.Sper.,50,1242-1247, 1974.
35.Stöcker,E.,Schultze,B., Heine,W.D. and Liebcher,H.Wachstum und Regeneration in parenchymatösen Organen der Ratte, Z.Zellforsch.,125,306-331,1972.
36.Stöcker,E. and Wullstein, H.K. Capacity of liver regeneration after partial hepatectomy in cirrhotic and CCl intoxicated old rats, In: Liver Regeneration after Experimental Injury, Freiburg Br., Ed. Lesch,R. and Reutter, W., 66-74, 1973.
37.Takata, T. Role of liver functions in liver cell mitosis, Acta Med.Okayama,28, 199-212, 1974.
38.Todorov, I.N., Kornyushina, T.V.,Govze, Yu.G. and Galkin, A.P. A study of ultrastructural and functional genome expression under prolonged suppression of protein biosythesis with cycloheximide, Communication V, Cytology and Genetics,1, 22-25, 1976, (In Russian).
39.Tsanev, R. Cell cycle and liver function. In: Results and Problems in Cell Differentiation, Springer - Verlag, Berlin-Heidelberg-New York, 7,197-248, 1975.
40.Tuczeck,H.V., Rabes,H. Verlust der Proliferations-fahigkeit der Hepatozyten nach subtotaler Hepatektomie, Experientia,27,526,1971.
41.Uryvaeva, I.V. and Faktor, V.M. Cell division - function relationship. Liver tolerance to toxic effect of CCl after partial hepatectomy,Cytology,18,1354-1358,1976 (In Russian).
42.Van Cantfort, I.I.,Barbason,H.R. Relation between the circadian rhythms of mitotic rate and cholesterol-7α-hydroxylase activity in the regenerating liver, Cell Tiss.Kinet., 5, 325-

330, 1972.
43. Verity, M.A., Travis, G. and Cheung, M. Lysosome-vacuolar system reactivity during early cell regeneration, Exper.Molec. Pathol., 22,73-84,1975.
44. Weinbren,K., Taghizadeh,A. The mitotic response after subtotal hepatectomy in the rat,Brit.J.Exper.Pathol.,46,413-417,1965.
45. Weinbren,K. and Woodward,E.Delayed incorporation of ^{32}P from orthophosphate into deoxyribonucleic acid of rat liver after subtotal hepatectomy, Brit.J.Exper.Pathol.,45,442-449,1964.
46. Yager, J.D., Hopkins,H.A., Campbell, H.A. and Van Potter, R. An autoradiographic analysis of DNA synthesis following partial hepatectomy in rats,In:Liver Regeneration after Experimental Injury, Freiburg Br., Eds. Lesch,R. and Reutter, W.,56-60. 1973.
47. Yakovlev, A.Yu. Kinetics of proliferative processes induced by phytohemagglutinin in irradiated lymphocytes, Radiobiology, 23,449-453, 1983 (In Russian).
48. Yakovlev, A.Yu., Malinin, A.M., Terskikh, V.V. and Makarova, G.F.Kinetics of induced cell proliferation at steady-state conditions of cell culture, Cytobiologie, 14, 279-283, 1977.
49. Yakovlev, A.Yu.and Zorin, A.V..Computer simulation in cell radiobiology,Springer-Verlag, Berlin, Heidelberg, New York, 1988.
50. Yakovlev A.Yu., Zorin, A.V. and Isanin, N.A. The kinetic analysis of induced DNA synthesis,J.Theor.Biol.,64, 1-25, 1977.
51. Zorin A.V., Yakovlev A.Yu. The properties of cell kinetics indicators. A computer simulation study,Biom.J.,28, 347-362, 1986.

CONCLUSION

At present there are no longer any doubts that adequate characterization of the kinetics of proliferative processes is impossible without employing methods of the present-day mathematical theory of cell systems and staging experiments designed specifically for the application of such methods. In solving concrete problems of cell population kinetics need arises at all times for widening the applicability limits of the existing methods of the cell system theory, as well as for bringing closer together the available theoretical results and accumulated evidence. In other words, it is necessary to consistently adapt the model in use to various known features of the organization of cell proliferation processes and actual potentialities of biological experiment.

As a rule, for the analysis of experimental findings theoretical results are used which are valid only for a steady exponential or in a strict-sense steady states of the population. In the applied analysis of cell proliferation kinetics the part played by transient processes is, indeed, underestimated. At the same time, it is separating out transient processes in the description of transitive cell population dynamics in the class of linear stationary (time-invariant) dynamic systems that leads to natural generalization of many cell kinetics methods and promotes their extension to a wider range of experimental phenomena. The results of studying the behaviour of the theoretical labelled mitoses curve in computer-aided experiments and of analyzing the diurnal rhythm of proliferative processes presented in this book (Chapter IV) show that consideration for the dynamics of transient processes is a "must" for obtaining satisfactory estimates of the parameters of mitotic cycle phase durations whenever there is no reason to regard as stationary the age distribution of cells in the cycle.

In reviewing different variants of non-stationary distribution of cell age in a life-cycle phase, the distribution of all phase durations was assumed to be time-independent. The experience in applying methods of the mathematical theory of cell systems to analysis of different experimental data shows that in many cases this assumption is quite plausible. However, investigation into the behaviour of cell kinetic indices undergoing the diurnal rhythm in the hamster cheek pouch (Chapter IV) has shown that allowances should be made for possible time-variations in the numerical parameters of distribution (temporal parameters) of mitotic cycle phase durations. This puts emphasis on the topical problem of further generalization of the cell kinetics model and imposes more stringent requirements on the accuracy of experimental data used in the identification of temporal cell cycle parameters. As pointed out in Chapter IV, work associated with the description of cell kinetics in the class of dynamic systems of time-variable structure (non-stationary systems) has been given much space in the literature.

Under the conditions of unsteady cell kinetics the traditional interpretation of experimental data in terms of temporal parameters of mitotic cycle phases yields insufficient information, and need arises for a comprehensive description of the flux of cells into the cycle phase under consideration. Such a description may be made using the q-index introduced in Chapter III which is not only of independent significance for analyzing the results of radioautographic experiment but is also instrumental in plotting theoretical labelled mitoses curves for different states of cell proliferation. Thus, combination in a single computation procedure of the methods for constructing the q-index for the S-phase of the mitotic cycle and the labelled mitoses curve makes it possible to obtain temporal parameter estimates for cycle phases in systems of unsteady cell kinetics and, at the same time, carry out more detailed analysis of such systems.

The advantages of introducing the q-index are particularly well demonstrated by the results of its application to the analysis of processes involved in induced cell proliferation. Some of the regular features of induced proliferation of cells, revealed by means of kinetic analysis and described in the present study, appear to be common to any of the cell systems capable of proliferative response to a stimulating effect. Among those are transition of cells to DNA synthesis in individual cell population fractions and the presence in the prereplicative period of an obligatory stage (transformation period) in passing through which a resting cell acquires proliferative competence. The existence of the transformation period in systems with induced proliferation of cells may be accounted for by the ability of cells "to sink" into the quiescent state which is, most probably, associated with their partial differentiation. On the other hand, the mean duration of the prereplicative period varies with the system and may change due to added damaging effects, such as irradiation. The phenomenon of "dynamic replacement" of cells described in Chapter V may also be considered as a feature of the system consisting of polyfunctional elements (cells) and being in the critical state as regards performance of its specialized functions, as is the case with the liver following an extensive resection of the hepatic parenchyma.

The above are but tentative considerations since there are a great number of experimental systems with induced or stimulated proliferation whose cell kinetics have not been adequately investigated.

Induction of proliferative processes in mammalian cells may be caused by a variety of means: in the liver - by partial hepatectomy, by administration of inducing mixtures (e.g., 3J-thyronine, glucagon, and the complete mixture of aminoacids),or agents damaging the hepatic parenchyma, and by different functional loads on cellular metabolism; in the kidney - by unilateral nephrectomy; in the intestinal epithelium - by nutritional loads in starving animals, by changing rations or intestine resection; in muscular fibres - by lesions leading to disintegration of the contractile substance; in the salivary gland - by administration of isoproterenol or its analogues; in the reproductive organs - by hormonal stimulation; in a monolayer cell culture - by changing the medium, adding fresh serum, certain enzymes or detergents, raising Ca^{++} ion content in a serum-free medium or mechanical injuring the monolayer; in a suspension culture of lympocytes - by adding phytohemagglutinin, concanavalin A or other mitogens.

In most animal tissues stimulation of cell proliferation is induced by damaging effects.Cell multiplication stimulators are represented by a great variety of labilizers of cell and lysosomal membranes, as well as substances damaging the microtubular and microfibrilic apparatus.

The list of systems with induced proliferation and of factors stimulating the proliferative activity of cells can be expanded considerably once we turn to the voluminous literature dealing with cytological and biochemical studies of cell multiplication mechanisms. However, the apparent deficiency of comprehensive data on cell population kinetics even in the most popular experimental systems perceptibly retards the advance in the exploration of specific molecular mechanisms inducing DNA synthesis and mitotic division. As a matter of fact, the results of biochemical investigation into systems with induced cell proliferation are useful for understanding such mechanisms only when they are closely related to a given level of cell maturity in the mitotic cycle. Were it possible to create standard conditions for conducting experiments which would ensure a synchronous entry into the mitotic cycle of an overwhelming proportion of a cell population, the problem of cytological interpretation of the results of biochemical research in proliferative processes would not be so pressing since it would suffice to take into account the regularities of cell population desynchronization. In reality, stimulation of cell proliferation not infrequently manifests itself in synchronized entry into the prereplicative period of but a minor fraction of cells, and many important metabolic changes occurring in those cells may prove to be beyond the sensitivity range of biochemical methods. Moreover, in in vivo systems a proliferative stimulus, in transferring a portion of the cell population into the mitotic cycle, may induce distinct changes, though not directly related to proliferation processes, in the biochemical properties of the other cells. In such cases it is not easy at all to assign the biochemical shifts observed in the experiment to a particular portion of the cell population.

This circumstance is rarely taken into account in interpreting experimental observations, while biochemical and molecular-biological investigations are far from always accompanied by concurrent experiments which would enable to evaluate peculiarities of cell population kinetics in each specific system. Under such conditions the role of mathematical and simulation methods oriented to the analysis of unsteady cell kinetics appears to be quite important.

The authors are hopeful the book will promote wide application of such methods in experimental practice.

SUBJECT INDEX

Lecture Notes in Biomathematics

Vol. 58: C. A. Macken, A. S. Perelson, Branching Processes Applied to Cell Surface Aggregation Phenomena. VIII, 122 pages. 1985.

Vol. 59: I. Nåsell, Hybrid Models of Tropical Infections. VI, 206 pages. 1985.

Vol. 60: Population Genetics in Forestry. Proceedings, 1984. Edited by H.-R. Gregorius. VI, 287 pages 1985.

Vol. 61: Resource Management. Proceedings, 1984. Edited by M. Mangel. V, 138 pages 1985.

Vol. 62: B. J. West, An Essay on the Importance of Being Nonlinear. VIII, 204 pages. 1985.

Vol. 63: Carme Torras i Genís, Temporal-Pattern Learning in Neural Models. VII, 227 pages, 1985.

Vol. 64: Peripheral Auditory Mechanisms. Proceedings, 1985. Edited by J. B. Allen, J. L. Hall, A. Hubbard, S. Neely and A. Tubis. VII, 400 pages. 1986.

Vol. 65: Immunology and Epidemiology. Proceedings, 1985. Edited by G. W. Hoffmann and T. Hraba. VIII, 242 pages. 1986.

Vol. 66: Nonlinear Oscillations in Biology and Chemistry. Proceedings, 1985. Edited by H. G. Othmer. VI, 289 pages. 1986.

Vol. 67: P. H. Todd, Intrinsic Geometry of Biological Surface Growth. IV, 128 pages. 1986.

Vol. 68: The Dynamics of Physiologically Structured Populations. Edited by J. A. J. Metz and O. Diekmann. XII, 511 pages. 1986.

Vol. 69: S. H. Strogatz, The Mathematical Structure of the Human Sleep-Wake Cycle. VIII, 239 pages. 1986.

Vol. 70: Stochastic Methods in Biology. Proceedings, 1985. Edited by M. Kimura, G. Kallianpur and T. Hida. VI, 229 pages. 1987.

Vol. 71: Mathematical Topics in Population Biology, Morphogenesis and Neurosciences. Proceedings, 1985. Edited by E. Teramoto and M. Yamaguti. IX, 348 pages. 1987.

Vol. 72: Modeling and Management of Resources under Uncertainty. Proceedings, 1985. Edited by T. L. Vincent, Y. Cohen, W. J. Grantham, G. P. Kirkwood, and J. M. Skowronski. VIII, 318 pages. 1987.

Vol. 73: Y. Cohen (Ed.), Applications of Control Theory in Ecology. Proceedings, 1986. VII, 101 pages. 1987.

Vol. 74: A. Y. Yakovlev, A. V. Zorin, Computer Simulation in Cell Radiobiology. VI, 133 pages. 1988.

Vol. 75: H. Knolle, Cell Kinetic Modelling and the Chemotherapy of Cancer. VIII, 157 pages. 1988.

Vol. 76: C. A. Macken, A. S. Perelson, Stem Cell Proliferation and Differentiation. VIII, 113 pages. 1988.

Vol. 77: A. Hastings (Ed.), Community Ecology. Proceedings, 1986. VII, 131 pages. 1988.

Vol. 78: J. J. Linderman, D. A. Lauffenburger, Receptor/Ligand Sorting Along the Endocytic Pathway. VI, 164 pages. 1989.

Vol. 79: P. Prusinkiewicz, J. Hanan, Lindenmayer Systems, Fractals, and Plants. V, 120 pages. 1989.

Vol. 80: D. House, Depth Perception in Frogs and Toads. VII, 135 pages. 1989.

Vol. 81: C. Castillo-Chavez, S. A. Levin, C. A. Shoemaker (Eds.), Mathematical Approaches to Problems in Resource Management and Epidemiology. Proceedings, 1987. VII, 327 pages.

Vol. 82: A. Yu. Yakovlev, N. M. Yanev, Transient Processes in Cell Proliferation Kinetics. VI, 214 pages. 1989.

This series reports new developments in biomathematics research and teaching – quickly, informally and at a high level. The type of material considered for publication includes:

1. Original papers and monographs
2. Lectures on a new field or presentations of new angles in a classical field
3. Seminar work-outs
4. Reports of meetings, provided they are
   a) of exceptional interest and
   b) devoted to a single topic.

Texts which are out of print but still in demand may also be considered if they fall within these categories.

The timeliness of a manuscript is more important than its form, which may be unfinished or tentative. Thus, in some instances, proofs may be merely outlined and results presented which have been or will later be published elsewhere. If possible, a subject index should be included. Publication of Lecture Notes is intended as a service to the international scientific community, in that a commercial publisher, Springer-Verlag, can offer a wide distribution of documents which would otherwise have a restricted readership. Once published and copyrighted, they can be documented in the scientific literature.

Manuscripts

Manuscripts should be no less than 100 and preferably no more than 500 pages in length.
They are reproduced by a photographic process and therefore must be typed with extreme care. Symbols not on the typewriter should be inserted by hand in indelible black ink. Corrections to the typescript should be made by pasting in the new text or painting out errors with white correction fluid. The typescript is reduced slightly in size during reproduction; best results will not be obtained unless on each page a typing area of 18 × 26.5 cm (7 × 10½ inches) is respected. On request the publisher can supply paper with the typing area outlined. Move detailed typing instructions are also available on request.

Manuscripts generated by a word-processor or computerized typesetting are in principle acceptable. However if the quality of this output differs significantly from that of a standard typewriter, then authors should contact Springer-Verlag at an early stage.

Authors of monographs and editors of proceedings receive 50 free copies.

Manuscripts should be sent to Prof. Simon Levin, Section of Ecology and Systematics, 345 Corson Hall, Cornell University, Ithaca, NY 14853-0239, USA, or directly to Springer-Verlag Heidelberg.

Springer-Verlag, Heidelberger Platz 3, D-1000 Berlin 33
Springer-Verlag, Tiergartenstraße 17, D-6900 Heidelberg 1
Springer-Verlag, 175 Fifth Avenue, New York, NY 10010/USA
Springer-Verlag, 37-3, Hongo 3-chome, Bunkyo-ku, Tokyo 113, Japan

---

ISBN 3-540-51831-2
ISBN 0-387-51831-2